Alternative materials in road construction

A guide to the use of waste, recycled materials and by-products

P.T. Sherwood

Published by Thomas Telford Publications, Thomas Telford Services Ltd, 1 Heron Quay, London E14 4JD

First published 1995
Reprinted 1997

Distributors for Thomas Telford books are
USA: American Society of Civil Engineers, Publications Sales Department, 345 East 47th Street, New York, NY 10017-2398
Japan: Maruzen Co Ltd, Book Department, 3—10 Nihonbashi 2-chome, Chuo-ku, Tokyo 103
Australia: DA Books and Journals, 648 Whitehorse Road, Mitcham 3132, Victoria

A catalogue record for this book is available from the British Library

Classification
Availability: Unrestricted
Content: Current best practice
Status: Established knowledge
User: Civil and transport engineers

ISBN: 0 7277 2018 X

Typeset in Times 10/11pt

Printed in Great Britain by Redwood Books, Trowbridge, Wilts.

Contents

GENERAL INTRODUCTION

Economic growth has inevitably led to increasing demands for aggregates for use in civil engineering construction. In the 30-year period to 1990 the total annual production within the UK of aggregates (sand, gravel and crushed rock) increased from 110 million tonnes to nearly 300 million tonnes (Fig. 1). Road building plays a significant role in this demand as it accounts for about one-third of the total production (Fig. 2). On average 20 000 tonnes of aggregate are used for each mile of motorway construction, and a total of 96 million tonnes of aggregate was used in road construction and maintenance in 1989. It is estimated that the current road building plans of the Department of Transport will use 510 million tonnes.

The environmental impacts of the extraction of aggregates are a source of significant concern across the country. These impacts include the loss of mature countryside, visual intrusion, heavy lorry traffic on unsuitable roads, noise, dust and blasting vibration. The extraction of aggregates also represents the loss of two finite natural resources: the aggregates themselves, and the unspoilt countryside from which they are extracted.

Concurrent with the production of aggregates, large amounts of waste materials and by-products are produced from industry and from domestic use. The relative amounts and types of waste produced are shown in Fig. 3, and a summary of the production of the mineral wastes (27% of the total), which form the subject of this book, is given in Table 1. Further details of the production and properties of these materials are given in later chapters.

If the materials listed in Table 1 could be used as alternatives to some of the materials used in civil engineering, their use would have the threefold benefit of conserving natural resources, disposing of the waste materials which are often the cause of unsightliness and dereliction, and clearing valuable land for other uses. This book therefore considers the extent to which waste materials and industrial by-products can be used in road construction in place of the natural materials that are traditionally used.

The book is divided into three parts. Part 1 discusses the demand and requirements of roadmaking materials and the specifications that they have to meet if they are to give satisfactory performance in each of the road pavement layers, from bulk fill at the bottom to the aggregates used in the surface layers at the top. Part 2 describes, in turn, each of the materials that are available as viable alternatives to the naturally-occurring materials that are traditionally used. It describes their physical and chemical properties and discusses their potential for use as roadmaking materials. Part 3 discusses the reasons why

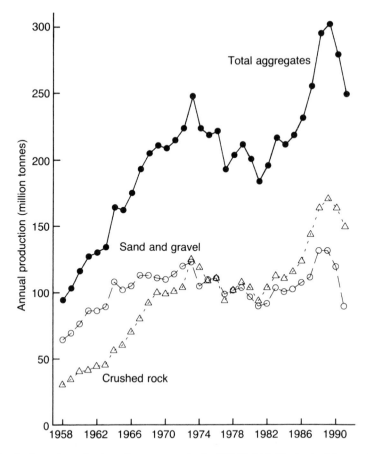

Fig. 1. Annual production of aggregates in the UK (British Geological Survey 1992)

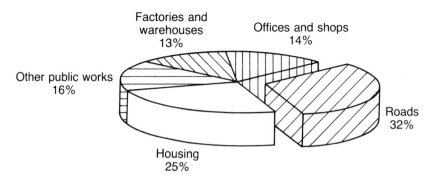

Fig. 2. Percentage of aggregates used in construction 1990 (BACMI 1991)

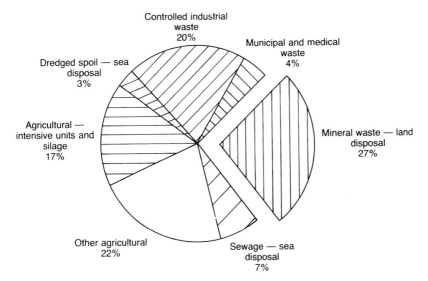

Fig. 3. Estimated annual wastes arising in the UK (Royal Commission on Environmental Pollution 1985)

Table 1. Availability of the major waste materials and by-products. (Whitbread et al. 1991)

Material	Production (million tonnes/year)	Stockpile (million tonnes)	Regions where available
Colliery spoil	45	3600	Coal mining areas
China clay wastes	27	300 (sand)	Cornwall and Devon
		300 (other)	Cornwall and Devon
Pulverized fuel ash	10	Some	Power stations
Furnace bottom ash	2·5	Some	Power stations
Blast furnace slag	5·8	Some old tips	North Yorkshire, Humberside, Wales
Steel slag	2	12	North Yorkshire, Humberside, Wales
Slate waste	4·5	450	North Wales, Lake District, South West England
Spent oil shale	0	150	Lothian region of Scotland
Road planings	7	—	Countrywide
Demolition wastes	24	—	Countrywide
Incinerated refuse	0·25		

it is desirable that waste materials and by-products should be used in preference to naturally-occurring materials. It examines the optimum use of wastes and by-products, bearing in mind the need to minimize resource depletion, environmental degradation and energy consumption.

PART 1
REQUIREMENTS FOR
ROADMAKING MATERIALS

If alternative materials are to be used in road construction they have to be classified and meet specification requirements in the same way that classification systems and specifications have been drawn up for roadmaking materials already in use. In theory, the specifications for alternative materials need not necessarily be the same as those for traditional materials, but in current practice this is invariably the case, although, as will be seen later, this may change. Before describing, in Part 2, the potential uses of the various alternative materials that are available, Part 1 therefore considers the requirements that they can be expected to meet if they are to be seriously considered.

1. Classification and sources

Classification of roadmaking materials

Roadmaking materials can be classified either by the type of material, e.g. aggregate, cement, bitumen, etc. or by the pavement layer in which they are to be used, e.g. sub-base material. In the second example, which is more appropriate to alternative materials, it is necessary, before discussing the demand for, and specification of, roadmaking materials, to consider the methods currently in use for the vast majority of road construction projects.

Layer structure of roads

A road structure is made up of a number of layers (pavement layers) which are shown in Figs 4 and 5. Two main types of construction are used:

(a)   flexible, in which the top layers are bituminous-bound (Fig. 4)
(b)   rigid, where the top layer is high-quality concrete (Fig. 5).

There is also a third category (not illustrated), known as composite construction,

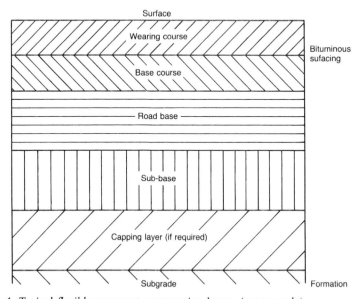

Fig. 4. Typical flexible pavement construction layers (not to scale)

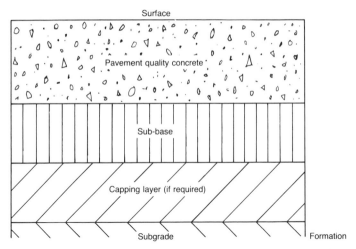

Fig. 5. Typical concrete pavement (rigid) construction layers (not to scale)

where the upper layers are constructed from bituminous material and are supported on a road base or lower road base of cement-bound material.

In a layered road pavement structure of the type considered here, the quality, in terms of durability and bearing capacity, of each of the pavement layers increases from the bottom upwards, i.e. the specification requirements for any given layer are always higher than those of the layer immediately beneath it. This means that the same material could be used for the construction of a particular layer and all the underlying layers; in principle, the whole road structure could be constructed from the materials used for the top layer. Full-depth asphalt construction goes some way to achieving this aim. However, building in layers generally means that costs are reduced and a very wide range of construction materials can be used. These materials range from material that occurs free on site for the construction of the bottom layer, to high-cost aggregates of high strength and skid resistance for the construction of the surface layer. This means that the scope for finding alternative materials as replacements for naturally-occurring materials decreases as the specification requirements for the respective layers increases. Bulk fill is therefore the biggest potential outlet for wastes, while very few such materials can be used in the surface layers.

The bottom of the road structure on which the pavement layers are constructed is known as the subgrade. This may be in-situ material (usually soil), or fill material which has been imported to make up the level or to replace the in-situ material if this is too unstable to permit construction to proceed. The subgrade plays an important role in the design of the road structure as its bearing capacity decides the thickness of the road structure above it. The bearing capacity of the subgrade is most frequently measured in terms of the California Bearing Ratio (CBR), which is an empirical test developed by the California State Highways Department for the evaluation of subgrade strengths. The procedure relates all materials back to a well-graded and non-cohesive

crushed rock which is considered to have a CBR value of 100. Examples of
how the CBR value is used in practice are given in Figs 6 and 7, which show
how the design thicknesses of the two layers immediately above the subgrade
(i.e. the capping layer and the sub-base) are dependent on the CBR of the
subgrade.

Bulk-fill materials

If the required level of a new road is not the same as that of the ground
over which it is built, the level has to be lowered or raised, processes known
as 'cut' and 'fill' respectively. As far as possible the road engineer will try

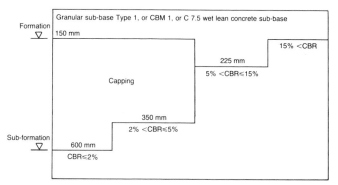

Granular sub-base Type 2 may be used as sub-base for design traffic loadings less than 400 commercial vehicles/day at opening
Granular sub-base Type 2 shall have a CBR of 30% or more when tested in accordance with clause 804.

*Fig. 6. Design thickness for flexible and composite pavement of capping and sub-base
for different CBR values of subgrade (Department of Transport 1987)*

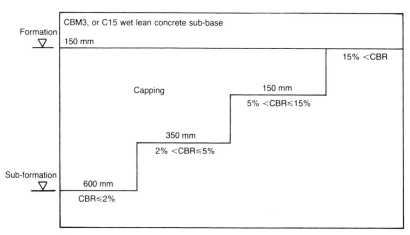

CBM2 or C10 wet lean concrete may be used at sub-base for design
traffic loadings less than 700 commercial vehicles/day at opening.

*Fig. 7. Design thickness for rigid and concrete pavement of capping and sub-base for
different CBR values of subgrade (Department of Transport 1987)*

to balance the volumes of cut and fill, but inevitably there will be occasions when extra material has to be imported to the site. Moreover, the exercise of balancing 'cut' and 'fill' is complicated by the fact that it is sometimes difficult, at the design stage, to identify the soils which will be suitable for excavation and re-compaction on site. The suitability of some materials, such as clayey soils, depends on moisture content and the way they are 'worked'.

Imported fill may be required in very large quantities. The major requirements of imported fill are that it should be relatively easy to transport, to place and to compact. Once compacted all that is required of it is that it shall provide a stable bed, strong enough to receive the layer above it which may, depending on circumstances, be the capping layer or the sub-base. These conditions are fairly easy to meet and it is not difficult to find suitable materials; for this reason bulk fill provides by far the biggest potential outlet for the use of alternative materials.

Aggregates

Aggregates are required at all levels of the pavement structure except for the subgrade (the ground on which the road structure is founded), although their use for imported fill is not precluded. It is also possible, in favourable circumstances, to dispense with aggregates for the construction of the capping layer and sub-base by upgrading the subgrade by stabilization with lime or cement.

Large volumes of aggregate are consumed by the road building programme — 96 million tonnes were used in 1989 (29% of total aggregate production) and the amount of aggregates used in road construction is likely to remain high, even if only to maintain the existing road network. Over and above the aggregates used in carriageway construction large quantities are often required for ancillary works. In rural schemes the most important ancillary works, from the view point of aggregate usage, are bridges, drains, kerbs and verges. In urban schemes, footways, subways and other non-carriageway works become equally, or more, important. Please and Pike (1968) estimated that for major road construction $1 \cdot 3$ tonnes of aggregate per square metre of road pavement were required, while the ancillary works imposed an extra demand on average of about half of this amount.

The sub-base, capping and imported fill represents, in terms of volume (but not of cost), by far the greatest proportion of the road structure. Figs 6 and 7 show that the combined thickness of capping and sub-base may be as much as 750 mm and is never less than 150 mm. Table 2 gives the estimated amounts of granular material used at capping layer and sub-base level for different thicknesses of construction.

To the aggregates that are used in the capping layer and sub-base must be added the materials used in earthwork and related construction. These are not classified as aggregates and do not therefore appear in the statistics, so it is difficult to give a precise estimate of what this proportion might be. These materials may occur on site, but they frequently have to be imported in large quantities and can in themselves make up a significant proportion of the total amount of material used in road construction.

In contrast to the combined thickness of the lower layers, the greatest design

Table 2. Estimated aggregate demand for various thicknesses of sub-base and capping of a three-lane motorway (Collins et al. 1993)

Thickness of capping* (mm)	Thickness of granular sub-base* (mm)	Thickness of capping and granular sub-base* (mm)	Weight of aggregate (tonnes/km)
0	150	150	13 000
0	225	225	20 000
350	150	500	43 000
600	150	750	65 000

* Thicknesses are those given in Figs 6 and 7

thickness of the most heavily-trafficked road would never exceed a total thickness of 450 mm of road base and surfacing. Thus, although the specification requirements for bulk fill, capping and sub-base are less onerous than those for the upper layers, these foundation layers are likely in represent the bulk of the material requirements for road pavement construction. It is clear from this that the total thickness of the sub-base, capping layer and any imported fill is likely to be at least half of the total thickness of the whole road structure.

Binders

Soils, aggregates and related materials are used in an unbound condition in the lower layers of the road pavement. However, as they do not have sufficient stability for use in the upper layers they invariably have to be mixed with a binding agent which bonds the particles together by physical and/or chemical means. The binding agents of particular importance are bitumen, Portland cement and, to a much lesser extent, lime (calcium oxide CaO; and calcium hydroxide $Ca(OH)_2$). The binder content is only a small proportion of a bound mixture so that, in comparison with the amounts of aggregates that are used in road construction (Figs 1 and 2), the amounts of binder used are relatively small, but still significant (Table 3).

Road construction accounts for most of the consumption of bitumen but less than one fifth of the consumption of Portland cement. The properties of both may be enhanced by the use of secondary additives but, in the case of bitumen, the scope for using any of the alternative materials considered in this book as an additive is almost non-existent. By contrast, considerable technical, economic and environmental benefits may be gained from the addition of pulverized fuel ash and blast furnace slag to Portland cement, and these are considered further in Parts 2 and 3.

Sources of roadmaking materials

Naturally-occurring materials

Naturally-occurring materials have traditionally been used for road construction and they still represent by far the biggest source of supply. The materials may occur on site, as in the case of the in-situ soil or, more typically, as in the case of sand, gravel and crushed rock, they may be quarried.

Table 3. Production of Portland cement and bitumen in the UK 1980–1990 (Annual Abstract of Statistics 1991)

Year	Bitumen produced for inland consumption (million tonnes)	Cement production
1980	Not given	14·8
1981	1·67	12·7
1982	1·96	13·0
1983	1·99	13·4
1984	1·90	13·5
1985	1·89	13·3
1986	2·02	13·4
1987	2·16	14·3
1988	2·34	16·5
1989	2·42	16·8
1990	2·49	14·7

Most specifications for roadmaking materials are consequently based on the assumption that natural materials will be used. These materials may be used as they occur in nature, or they may be processed to produce a more consistent product, or they may be upgraded by the addition of a binding agent such as cement or bitumen.

Manufactured materials

Manufactured materials in this context refers to products that do not occur naturally but are specifically manufactured for use in building and civil engineering construction; they are distinct from those that are produced as by-products or wastes from another manufacturing process. Portland cement, bitumen and lime are the chief examples of manufactured materials used in significant quantities in road construction, and these have already been considered.

Apart from these, significant amounts of artificial aggregates are manufactured for use in civil engineering, for example to provide durable materials with light weight or good insulating properties. Compared with natural aggregates, these materials are expensive and their use is justified only for those purposes where the benefits of using them outweigh the considerable increase in costs. For example, in high-rise buildings, concrete made from lightweight aggregate reduces the building weight with consequent savings in the size of the foundations.

However, except possibly in the case of bridge construction and the production of materials with high skid resistance, there is little scope for using artificial aggregates in road construction.

Alternative materials (wastes and by-products)

Concern about the depletion of natural resources and the effect that meeting the demand for aggregates may have on the environment has, particularly in recent years, focussed attention on the possibility of finding alternatives to naturally-occurring materials. A number of possible sources of supply have been examined, the most important being the potential for use in road construction of the mineral waste materials and industrial by-products that are listed in Table 1. Other alternatives include the possibility of in-situ recycling of the road payment layers and the manufacture of artificial aggregates with properties similar to, or even superior to, natural aggregates.

Part 2 of this book discussed the possibilities of using alternative materials, and in Part 3 the environmental benefits of using them in preference to naturally-occurring materials are discussed. However, before this can be done the specifications in use for roadmaking materials need first to be examined, and this is the subject of the next chapter.

2. Specifications for roadmaking materials

British specifications

In the UK the requirements given in the Specification for Highway Works (1991a) are mandatory for all road construction which is funded by central Government. This is a national specification because, although usually referred to as the DOT Specification, it is issued jointly by the Department of Transport, the Scottish Office, the Welsh Office and the Department of the Environment for Northern Ireland.

A survey of the specifications of aggregates and bulk construction materials (Collins *et al.*) showed that the Specification for Highway Works formed the basis of most other specifications in use in the UK for road and related construction. It also showed that where departures were made from the national specification this was more likely to lead to more stringent requirements being made rather than to any relaxations. The review concluded that the national specification allowed a wide range of materials to be used and that it could not be regarded as being unduly restrictive.

The evaluation of a material for any particular application in road construction is therefore most readily done by comparing its properties with those of materials known to be satisfactory, as described in the national specification. Given that the specification is so widely used as the basis for all road construction in the UK there is, at present, little alternative to this approach. However, it suffers from the fact that most of the specifications in use are recipe-type, i.e. the properties of a material to be used for a given purpose are closely defined and it is assumed that if a material meets all the specification requirements it will perform satisfactorily. The material properties that should be specified are based on years of experience that materials with such properties will perform satisfactorily. This method works well with traditional materials, but when new materials are introduced they may fail to meet the specification requirements even if they would in fact be perfectly satisfactory.

Research is in progress to define more closely the requirements of the individual pavement layers so that 'end-product' specifications can be used which would permit the use of any material that could meet the design requirements of a particular pavement layer. However, until the results of this research are incorporated into new specifications there is no solution other than to accept that if alternative materials are to be used it must be shown that they can meet the existing specifications. However, for minor roads judicious relaxation of the requirements given in the national specification is a possibility that could lead to greater use of marginal materials which do not fully comply.

Certain of the more commonly occurring waste materials are mentioned by name in the national specification as being suitable or unsuitable for particular applications. Where this is the case the reasons for the inclusion or exclusion are discussed in the relevant part of this book. However, materials not mentioned may also be suitable or give rise to problems, and these cases are also considered.

Specifications for fill materials

The Specification for Highway Works (1991a) gives a wide range of fill materials which are regarded as acceptable, and defines unacceptable materials as:

(a)   material from peat, swamps, marshes and bogs
(b)   logs, stumps and perishable material
(c)   materials in a frozen condition
(d)   clays having a liquid limit in excess of 90% or a plasticity index exceeding 65% (note: only a small minority of clays would fall into this category)
(e)   material susceptible to spontaneous combustion, except unburnt colliery spoil compacted in accordance with with methods specified
(f)   material with hazardous chemical or physical properties.

A minor additional requirement is that when the fill is to be placed within 500 mm of concrete, cement-bound or within other cementitious materials, the soluble sulphate content should not exceed $1 \cdot 9$ g/litre. Similarly, if the fill is to be placed within 500 mm of a metal structure, the total sulphate content should not exceed $0 \cdot 5\%$.

Apart from common fill, the specification also includes requirements for special fills where aggregates are required. These fills are listed in the specification as selected granular fill (Class 6, see Table 4). Table 4 shows that the general requirements are for 'natural gravel, natural sand, crushed gravel, crushed rock, crushed concrete, chalk, well-burnt colliery spoil or any combination thereof'. Limits for these in terms of grading, plasticity, organic matter content and sulphate content and particle strength are specified.

Specifications for capping layers

The functions of a capping layer are:

(a)   to protect the subgrade from the adverse effects of wet weather
(b)   to provide a working platform on which the sub-base construction can proceed with minimum interruption in wet weather
(c)   to allow the full load-spreading capabilities of the sub-base to be realized, which would not be possible were it to be laid directly on top of a weak subgrade.

In recent years, the Department of Transport has encouraged the stabilization of the in-situ subgrade material for the construction of capping layers (Wood 1988).

Table 4. Requirements for unbound selected granular fills (Specification for Highway Works 1991a)

Qualifying term		Typical use	Permitted constituents	Principal requirements
6A	well graded	Below water	1,2,3,5,6	a,b,c
6B	coarse	Starter layer	1,2,3,4,5,6	a,b,c,e
6C	uniformly graded	Starter layer	1,2,3,5,6	a,b,c,e
6D	uniformly graded	Starter layer below PFA	1,2,5,6	a,b,c
6E	granular material	For stabilization	1,2,3,4,5,6	See Fig. 8
6F1	fine grading	Capping	1,2,3,4,5,6	See Fig. 8
6F2	coarse grading	Capping	1.2.3.4.5.6	See Fig. 8
6G	uniformly graded	Gabion filling	1,2,6	a,e
6H	uniformly graded	Drainage layer to Reinforced Earth	1,2,3,4,6	a,b,c,d,e
6I	well graded	Fill to Reinforced Earth	1,2,3,4,6	a,b,c,d
6J	uniformly graded	Fill to Reinforced Earth	1,2,3,4,6	a,b,c,d
6K	well graded	Lower bed to steel structures	1,2,5,6	a,b,c,d
6L	uniformly graded	Upper bed to steel structures	1,2,5,6	a,b,c
6M	well graded	Surround to steel structures	1,2,5,6	a,b,c,d,e
6N	well graded	Fill to structures	1,2,4,5,6	a,b,e
6P	uniformly graded	Fill to structures	1,2,3,4,5,6	a,b,e
6Q	granular material	Overlaying fill for buried metal structures	1,2,3	a,b,c,d,e

Key to permitted constituents:
1, Sand and gravel
2, Crushed rock
3, Chalk
4, Blast-furnace slag

5, Well-burnt colliery shale
6, Crushed concrete

Key to principal requirements:
a, Grading
b, Coefficient of uniformity
c, Plasticity
d, pH, sulphates, chlorides (only if metallic elements present)
e, 10% fines value (TFV)

The alternative to stabilization of the in-situ subgrade is either to use unbound granular material or to stabilize with cement imported granular material which would be unsuitable for use in an unbound form. The specification recognizes the following granular materials for capping.

Class 6E: Selected granular material for stabilization with cement, which may be any material, or combination of materials, other than unburnt colliery spoil and argillaceous rock.

Class 6F1: Selected granular material (fine grading) which may be any material, or combination of materials, other than unburnt colliery spoil.

Class 6F2: Selected granular material (coarse grading) which may be the same type of materials as are specified for Class 6F1.

The requirements for these materials are summarized in Fig. 8. Although it is not written into the specification, it can be inferred from other publications that the materials when placed should have a minimum CBR value of 15%. The CBR value is not an onerous requirement for either the unbound or cement-stabilized granular capping, and it can be seen from Fig. 8 that the grading requirements for unbound capping are very undemanding. Even if the grading requirements cannot be met a material may still be used if it satisfies the more relaxed grading requirement for cement-stabilized granular capping.

Specifications for sub-bases
The functions of a sub-base are:

(a)   to provide a working platform on which the paving materials can be transported and compacted
(b)   to be a structural layer which assists in spreading the wheel loads so that the subgrade is not overstressed
(c)   to be an insulating layer against freezing where the subgrade is a material likely to be weakened by frost action (to fulfil this function the sub-base must itself be frost-resistant).

The materials permitted in the UK for sub-base construction and the thicknesses required are given in Figs 6 and 7. These show that the thickness of the sub-base is not related to traffic intensity, and only to a slight extent on the bearing capacity of the subgrade; deficiencies in this are compensated for by requiring a greater thickness of capping layer. Figs 6 and 7 show that at sub-base level both unbound and cement-bound aggregates may be used. The specifications for these are discussed below.

Unbound sub-base and base materials. Unbound aggregates are described in clauses 803 and 804 of the specification which give, respectively, the requirements for granular sub-base material Type 1 and granular sub-base material Type 2. Until recently there was an additional clause for wet-mix macadam but this is little used and it has now been excluded from the specification. Type 1 granular materials include 'crushed rock, crushed slag, crushed concrete or well-burnt non-plastic shale'. Type 2 granular materials include the materials permitted for type 1 and also natural sands and gravels. Wet-mix macadam was restricted to crushed rock or crushed slag. The major requirements of Type 1 and Type 2 granular sub-base material are summarized in Fig. 9.

In addition to the requirements given in Fig. 9, unbound materials used within 450 mm of the road surface also have to be non-frost-susceptible when tested by the BS frost-heave test (BS 812 1988).

Type 1 materials, which exclude the use of sands and gravels, are expected to be all-weather sub-base aggregates; Type 2 materials have a lower specification and are not expected to give good performance under construction in wet weather. Fig. 7 shows that unbound sub-bases are not permitted in rigid pavement construction; they may be used in flexible pavement construction,

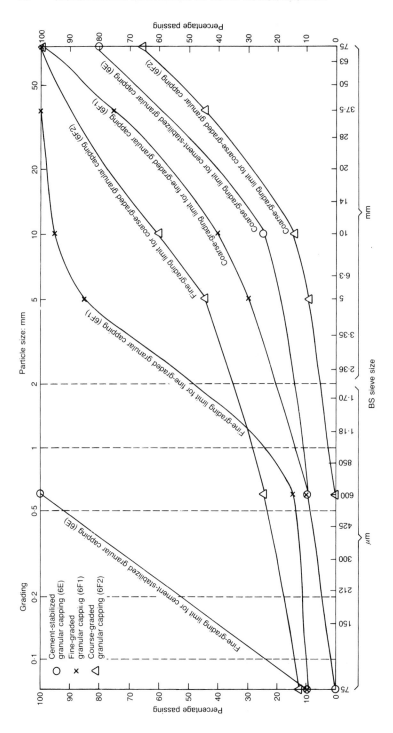

Fig. 8. Requirements for granular materials used in capping layer construction (Specification for Highway Works 1991)

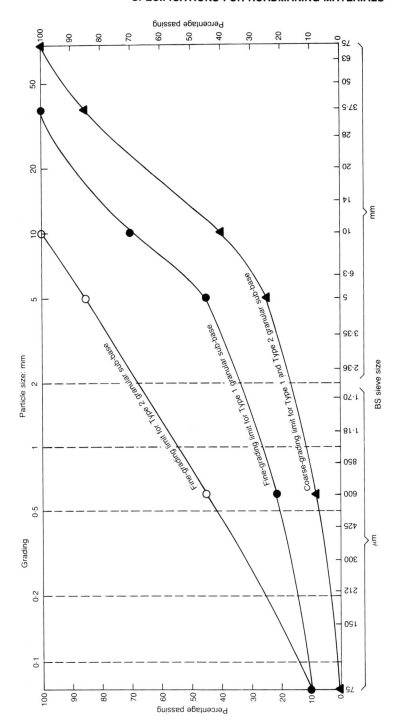

Fig. 9. Requirements for unbound granular materials used in sub-base construction (Specification for Highway Works 1991)

Other requirements	Class of material		
	6E	6F1	6F2
Liquid limit (%)	<45	NR	NR
Plasticity index (%)	<20	NR	NR
Organic content (%)	<2	NR	NR
Total sulphate (%)	<1	NR	NR
10% fines value (kN)	NR	>30	*

NR, no requirement specified
* Value written into the contract document

(Fig. 8. Cntd)

Other requirements	Type 1	Type 2
Plasticity index (%)	0	<6
Soaked 10% fines value (kN)	50	50
Soundness value (%)	>75	>75
Water absorption (%)	<2	<2
Minimum CBR	*	30

* CBR assumed to be adequate

(Fig. 9. Cntd)

but Type 2 materials are excluded if the design traffic loadings are more than 400 commercial vehicles per day.

Cement-bound sub-bases under flexible construction.   Figures 6 and 7 show that cement-bound materials may be used as an alternative to unbound granular materials under flexible construction and are the only permitted material under rigid construction. The specification contains clauses for a family of cement-bound materials referred to by their initials as CBM1, CBM2, CBM3 and CBM4 (these were previously known, respectively, as soil-cement, cement-bound granular material and lean concrete). All could be used at sub-base level but the superior quality of CBM3 and CBM4 means that they are used principally for base construction or for sub-base under rigid construction while CBM1 and CBM2, described respectively in clauses 1036 and 1037, are confined to sub-base construction.

The materials permitted for use for CBM1 and CBM2 are not specified as such, it being assumed that as long as they satisfy all the requirements of the specification they may be regarded as suitable. This means that, potentially, a very wide range of materials may be used. The materials to be used for the superior quality CBM3 and CBM4 are specified as natural aggregates complying with the requirements of BS 882 (1992), air-cooled blast-furnace slag complying with the requirements of BS 1047 (1983a), or crushed concrete

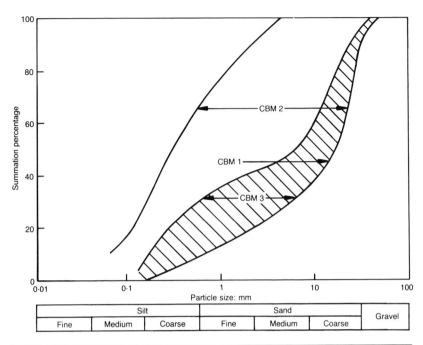

Other requirements	CBM1	CBM2	CBM3	CBM4
Soaked 10% fines value (kN)	NR	50	NR	NR
Average 7-day compressive strength (MPa)	4·5	7·0	10·0	15·0
Minimum 7-day compressive strength (MPa)	2·5	4·5	6·5	10·0
Strength of 7-day air-cured and 7-day soaked as percentage of 14-day air-cured	80	80	NA	NA

NR, no requirement; NA, not applicable

Fig. 10. Requirements for cement-bound sub-base and base materials (The grading requirements for CBM1 relate to the coarse side of the grading envelope; there is no fine limit. The requirements for CBM3 and CBM4 are the same) (Specification For Highway Works 1991)

which complies with the quality and grading requirements of BS 882 (1992). The requirements of the CBM group of cement-bound sub-base and base materials which have to be satisfied are summarized in Fig. 10.

As an alternative to the CBM group of materials, which are compacted by rollers, the specification also includes a range of four cement-bound sub-base and base materials known as 'wet lean concrete 1-4' which are compacted by vibratory methods.

Cement-bound sub-bases under rigid construction. Figure 7 shows that unbound sub-bases are not permitted to be used under rigid (concrete) construction, and the weaker cement-bound and wet lean concretes are also excluded. There are good reasons for this as the concrete paving is laid directly on to the sub-base (see Fig. 5) so that in effect the distinction between sub-base and base, which applies to flexible construction, does not exist. The sub-base/base has to provide a rigid platform on which the high quality concrete paving can be laid. The requirements cannot therefore be regarded as restrictive as it would be a false economy to risk the failure of expensive concrete paving by taking risks with the quality of the material on which it was laid.

Specifications for base construction

The road base must not be confused with the base course, which is an integral part of the surface course. The base course is a sub-layer within the bituminous surfacing; the road base is normally the thickest element of the flexible pavement on which the surfacing rests.

From a structural aspect, the road base is the most important layer of a flexible pavement. It is expected to bear the burden of distributing the applied surface loads so that the bearing capacity of the subgrade is not exceeded. Since it provides the pavement with added stiffness and resistance to fatigue, as well as contributing to overall thickness, the material used in a road base must always be of reasonably high quality. For this reason the scope for using alternatives to the naturally-occurring materials is severely restricted. The requirements of road base materials are not therefore considered here but are considered in Part 2, which deals with specific alternatives when these alternatives have some potential for use in road base construction.

Specifications for surfacing

The surfacing, and particularly the wearing course, represents only a small proportion of the total depth of construction. However, it accounts for a high proportion of the total cost of the pavement. Specifications for surfacing therefore impose more stringent requirements than is the case for the lower layers of the road pavement. The potential for using alternative materials in the surfacing is therefore very restricted and is considered, as appropriate, in Part 2.

Other national specifications

A recent survey of specifications in use for aggregates (Collins *et al.* 1993) showed that the specification of roadmaking aggregates in other countries did not differ significantly in fundamental principles from those used in this country. All countries adopt a layer technique (Figs 4 and 5) for road construction and use similar (but not identical) tests to determine suitability. This section reviews some of the national specifications that are in use.

Specifications in the USA

Unbound bases and sub-bases. Although each state in the USA is a completely independent highway authority, the American Association of State Highway and Transportation Officials (AASHTO) and the American Society for Testing Materials (ASTM) have both issued specifications for unbound sub-base and base materials.

AASHTO M147-65 specifies six types of aggregate, designated A−F. These

Table 5. AASHTO requirements for unbound sub-bases and base materials (AASHTO M 147-65, 1980)

Sieve size	Grading percentage passing					
	A	B	C	D	E	F
50 mm	100	100	100	100	100	100
25 mm	NR	75−95	100	100	100	100
9·5 mm	30−60	40−75	50−85	60−100	100	100
4·75 mm	25−55	30−60	35−65	50−85	55−100	70−100
2 mm	15−40	20−45	25−50	40−70	40−100	55−100
425 μm	8−20	15−30	15−30	25−45	20−50	30−70
75 μm	2−8	5−20	5−15	5−20	6−20	8−25

Other requirements:
liquid limit of < 425 μm fraction not greater than 25%;
plasticity index of < 425 μm fraction not greater than 6%;
percentage wear by Los Angeles test not greater than 50%.

differ with respect to their gradings (Table 5). Grading A is used primarily for bases and gradings B-D refer to sub-base materials. Gradings E and F are used as top courses for unsurfaced roads and have no counterparts in the UK. The requirements for plasticity and strength, as measured by the Los Angeles test, are also given in Table 5. Material A has a grading very similar to that of the British Type 1 sub-base material, and the extremes of the grading envelopes of materials B, C and D correspond quite closely to Type 2 sub-base material (see Fig. 9).

Two gradings, one for sub-base and the other for base, are specified by ASTM (D2940-74, 1985) and, with other requirements specified by ASTM, are given in Table 6. Comparison of these requirements with those of their

Table 6. ASTM requirements for unbound sub-bases and base materials (ASTM D2940-74, 1985)

Sieve size	Grading percentage passing	
	Bases	Sub-bases
50 mm	100	100
37·5 mm	95−100	90−100
19 mm	70−92	NR
9·5 mm	50−70	NR
4·75 mm	35−55	30−60
600 μm	12−25	NR
75 μm	0−8	0−12

Other requirements:
coarse aggregate to be hard, clean and durable;
fraction passing the 75 μm sieve not to exceed 60% of the fraction passing the 600 μm sieve;
liquid limit of < 425 μm fraction not greater than 25%;
plasticity index of < 425 μm fraction not greater than 4%.

British counterparts, given in Fig. 9, shows that the sub-base grading is more restrictive, particularly with regard to the finer grading limit, than Type 2 sub-base, but the grading for base material is very similar to the British Type 1 granular sub-base.

Many states incorporate some of the AASHTO and ASTM requirements into their specifications. However, with the large variations in geology and climate, the individual specifications used by the independent state highway authorities differ significantly among the states.

Geological differences play an important role: some states, such as Georgia and North Carolina, have abundant supplies of crushed rock and therefore prohibit the use of gravels. Others, with extensive gravel deposits, do not restrict the use of gravel. In general, however, where gravel is permitted most states have a requirement that all gravel particles must have at least one face fractured by crushing.

All the states give a grading requirement and also have requirements for durability and the amount of plastic fines. The Los Angeles test is most widely used for specifying durability; restrictions on the amount of plastic fines are imposed by specifying a maximum plasticity index or sand equivalent value.

Cement-bound bases and sub-bases. As in the case of unbound materials, each state has its own specification for cement-stabilized material used in the different layers of the road pavement but, in so far as there is a national specification, the comprehensive recommendations published by the Portland Cement Association (PCA) (1971, 1977 and 1979) apply. The philosophy adopted by the PCA is that a hardened stabilized material should be able to withstand exposure to the elements. The types of material to be used and their grading are expressed in very broad terms; strength is considered to be a secondary requirement as most stabilized mixtures that possess adequate resistance to the elements also possess adequate strength.

The tests used to determine performance are ASTM wetting and drying test (ASTM D558) and the ASTM freezing and thawing test (ASTM D660). These are designed to simulate what happens in practice, and provided a stabilized material can meet the requirements of both tests it is considered to be suitable for use regardless of its origins. In this respect the philosophy is not unlike that of the British cement-bound material category 1 (CBM1) where few requirements for the material to be used are given.

European specifications

In the early 1980s the Organisation for Economic Co-Operation and Development (OECD) started a review among member countries of the use of marginal aggregates in road construction. This work was never completed but, in the course of the preparation of a draft report, delegates from each country (which included most of Western Europe plus Australia and Canada) were asked to provide information on the specifications used for aggregates in their respective countries. Table 7, which is taken from the draft report (OECD 1981), summarizes the information that was obtained.

With regard to unbound aggregates, Table 7 shows that there is usually (but not always — France being a notable exception) a grading requirement, a

Table 7. Comparison of the specifications for aggregates used in various OECD member countries

	Unbound sub-base	Stabilized base	Bituminous Surfacing Base course	Wearing course
Australia:				
Grading	—	Yes	Yes	Yes
Los Angeles	<35−50	<30−45	<25−40	<25−40
Plasticity index	Yes	Yes	—	—
Wet 10% fines value/ dry 10% fines value (%)	<35	<35	<35	<35
Polished stone value (PSV)	—	—	—	<43−47
Belgium:				
Grading	Yes	—	Yes	Yes
Particles <74 μm	<5	<2	<2	<2
Organic matter	<0·5	<0·5	<0·5	<0·5
PSV	—	—	—	>50
Canada:				
Grading	Yes	—	Yes	Yes
Petrographic no.	<200	—	<140	<130
Los Angeles	<40	—	<40	<40
Resistance to wear	<20	—	—	—
Organic matter	<0·8	—	—	—
Particles <74 μm	—	—	<1·5	<1·5
Denmark:				
Grading	Yes	Yes	Yes	Yes
Sand equivalent	>30	>30	—	—
Finland:				
Grading	Yes	Yes	Yes	Yes
Los Angeles	—	—	≤35	≤30
Dynamic fragmentation	—	—	≤70	≤60
Proportion of crushed particles	—	—	≥30%	≥30%
France:				
Crushing index	≥60	≥100	—	—
Los Angeles	≤25	≤30	≤25	≤15
MicroDeval (wet)	—	≤25	≤20	≤15
Particles <74 μm	—	—	≤2	≤2
Sand equivalent	≥40	≥40	≥50	≥50
PSV	—	—	—	≥50
Germany:				
Grading	Yes	Yes	Yes	Yes
Water absorption	<0·5	<0·5	<0·5	<0·5
Mechanical resistance	—	<34	<18	<18

Table 7. Cntd

	Unbound sub-base	Stabilized base	Bituminous Surfacing	
			Base course	Wearing course
Ireland:				
Grading	Yes	Yes	Yes	Yes
Plasticity index	0–6	—	—	—
10% fines value	≥ 50 kN	≥ 50 kN	—	—
Aggregate abrasion value (AAV)	—	—	—	< 12
Italy:				
Grading	Yes	Yes	Yes	Yes
Los Angeles	< 30	< 30	< 25	< 20
Sand equivalent	> 50	> 50	—	—
Immersion coefficient	—	—	< 0·015	< 0·015
Netherlands:				
Grading	Yes	—	Yes	Yes
Loss on ignition	< 3%	—	—	—
Compressive strength at 28 days	—	≥ 2 MPa	—	—
PSV	—	—	—	48–65
Spain:				
Grading	Yes	Yes	Yes	Yes
Crushing index	≥ 50	≥ 50	≥ 75	≥ 75
Los Angeles	< 35	< 30	< 25	< 25
Sand equivalent	≥ 30	≥ 30	≥ 45	≥ 45
Organic content	—	< 0·05	—	—
Argillaceous materials	—	< 2	—	—
Sulphates	—	< 0·5	—	—
PSV	—	—	—	> 45
Sweden:				
Grading	Yes	Yes	Yes	Yes
Argillaceous materials	< 5	—	—	—
Organic matter	< 2	—	—	—
Switzerland:				
Grading	Yes	Yes	Yes	Yes
Crushed particles (%)	—	—	> 40	100

strength requirement and a requirement to restrict the amount of plastic fines. The tests used to define these properties differ from country to country and it should not be assumed that the tests, even when they have the same name, necessarily give comparable results as there may be subtle differences in the test procedures in use in the different countries.

Future prospects for European specifications. In preparation for the single European market, the Comité Européen de Normalisation (CEN — European Committee for Standardisation) has, in recent years, been preparing European Standards which will cover the use of all materials used in road construction, including the use of waste materials and by-products. CEN's ultimate objective is to produce fully harmonized European Standards, denoted by EN, which will replace existing national standards, such as those produced by the British Standards Institution. Approval of draft European Standards is carried out by a formal vote among the member countries. If the necessary overall majority is obtained, after counting weighted votes, all countries are obliged to adopt the EN to replace their national standards regardless of their vote.

In addition to the ENs, which correspond to BS specifications, there is a set of pre-standards which correspond to BS Drafts for Development. If a pre-standard gains an overall majority vote, it is not necessary for conflicting national standards to be withdrawn and they may operate in parallel with the pre-standard. Like BS Drafts for Development, pre-standards have a finite life and are subject to reconsideration after a period of three years, when they can either be upgraded to full EN status or continue, in existing or modified form, for two years before further consideration.

As in the case of the British Standards Institution, CEN operates through a series of Technical Committees (TCs) and Sub-Committees (SCs) which in turn may appoint Working Groups (WGs) and Task Groups (TGs). Each member country is represented on the Technical Committees and may, if it wishes, be represented on the Sub-Committees, but membership of Working Groups and Task Groups is usually confined to individual experts in the particular area of interest.

Within the field covered by the materials included in this book there are three relevant Technical Committees.

(a)   TC 154 'Aggregates' which was established in late 1988. This is a 'Product Committee' which will describe aggregate properties by reference to test methods and will deal with aggregate particles. Aggregates in this context will include artificial aggregates, waste materials and by-products.

(b)   CEN TC 227 'Road Construction and Maintenance Materials', which was established in 1990. This is a 'Functional Committee' and operates to produce mixtures of ingredients (e.g. aggregates, cement, etc., as defined by a 'Product Committee') to fulfil an appropriate function, in this case the structure of a road, airfield or other paved area.

(c)   CEN TC 292 'Characterisation of waste products', a new TC with a wide scope and intended to include wastes used in construction. It was established in 1992 and has yet to define its method of operation.

Each of these Technical Committees has a number of Sub-Committees, Working Groups and Task Groups. TC 154 has six such Sub-Committees, two of which are of little relevance to the use of waste materials in road construction but which also include:

• TC 154 SC2 'Aggregates for concrete'
• TC 154 SC3 'Bituminous-bound aggregates'
• TC 154 SC4 'Hydraulic-bound and unbound aggregates'
• TC 154 SC6 'Test methods'.

TC 227 has a Working Group — WG4 'Unbound, hydraulic-bound, waste and marginal materials', which specifically includes waste materials within its terms of reference.

As the titles of TC 154 SC4 and TC 227 WG4 suggest, there is a fine dividing line between them, but this can be clarified by applying the principle that TC 227 refers to a product with a nominated end use, whereas TC 154 applies to aggregates with only a very general indication of end use.

None of the Committees mentioned has yet produced a published standard but the following drafts have been prepared.

(a)   TC 154 SC 4 has prepared a working draft of a proposed EN for 'Hydraulic bound and unbound aggregates for civil engineering work and road construction'. This is structured into three sectors:

(i)   intrinsic characteristics
(ii)   manufactured characteristics
(iii) additional characteristics.

Much work is still required before a final draft can be produced and progress is dependent on the development of appropriate test methods by TC 154 SC6.

(b)   TC 227 WG4 has prepared draft specifications for:

(i)   unbound mixtures of natural and artificial aggregates
(ii)   lime-bound mixtures for road capping layers
(iii) soil cement
(iv)   cement-bound granular materials
(v)   roller-compacted concrete.

The last three materials above correspond approximately to the three categories (CBM1, CBM2 and CBM3, respectively) of cement-bound material mentioned in the UK Specification for Highway Works (1991a). It is of interest to note that the terminology for cement-bound materials has largely reverted to the terminology used in pre-1986 editions of the specification.

PART 2
ALTERNATIVE MATERIALS AVAILABLE — QUANTITIES, LOCATIONS, GENERAL PROPERTIES AND POTENTIAL USES

This Part considers the alternative materials (wastes and by-products) that are available in the UK and which have some potential for use in road construction. Each of the major materials listed in Table 1 is considered in a separate chapter of this Part. Each chapter describes the sources, the quantities currently produced, and the quantities available from past production of the particular material. The chapters go on to consider the physical and chemical properties of the relevant material, and end with a discussion of its potential uses in road construction.

3. Colliery spoil

Occurrence

Colliery spoil is available in the coal mining areas of the UK (see Fig. 11) in very large quantities. Of the wastes and industrial by-products available, colliery spoil is currently produced in the largest amounts with an estimated annual production of 45 million tonnes (Table 8) available for disposal on land in 1988−89. To this must be added the estimated 3600 million tonnes available in the stockpile of spoil arising from past production.

The introduction, in the 1950s, of mechanized methods of working increased the proportion of spoil to coal, so spoil production greatly increased between 1950 and 1970 (Table 8). With the decline in coal production since then, the amount of spoil has decreased and can be expected to decrease further. However, the amounts produced are still likely to mean that colliery spoil will remain a major source of waste and, with huge stockpiles available from past production, it is by far the most significant of the materials potentially available for use in road construction.

Most of the spoil produced is tipped on land; in 1978 this amounted to about 50 million tonnes (88% of the then total production), of which 4 million tonnes were used in local land reclamation, with the rest forming spoil tips (Commission on Energy and the Environment 1981). About 7 million tonnes (12%) were disposed of using marine disposal methods, the majority being tipped on the foreshore rather than into deep water. About 0·5 million tonnes were removed from existing tips for commercial use.

Composition

Colliery spoil deposits are composed of the waste products from coal mining which are either removed to gain and maintain access to the coal faces or are unavoidably brought out of the pit with the coal and have to be separated at the coal cleaning plant. Wastes from both sources are usually dumped on the same spoil heaps, which are often referred to as 'shale' or 'slag' heaps. These heaps are a prominent feature of coal mining areas (see Fig. 12). (None of the materials should be regarded as slag, the definition of which, in civil engineering construction, should be confined to materials derived from the extraction of metals from their ores — see section on slags. Shale is also a misnomer as the tips contain materials other than shale).

Due to the manner in which they are formed the spoil tips are highly variable in composition. Superimposed is an additional variation arising from combustion in the heaps. When combustion occurs the physical and chemical

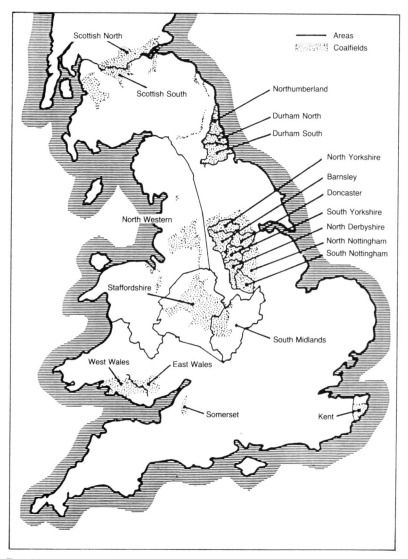

Fig. 11. Distribution of colliery spoil in the UK

composition is changed, burnt spoil (burnt shale) differs considerably in its properties from the unburnt spoil (minestone). The availability of burnt spoil is declining as, following the Aberfan disaster of 1966, great care is now taken in the construction of the spoil tips. Taken together with improvements in coal separation this means that the possibility of spontaneous combustion in modern spoil tips is negligible. Burnt spoil is therefore only available from older tips and is also in higher demand than the unburnt material (minestone), because it can be used in road sub-base and base construction. Supplies are therefore being rapidly depleted.

Table 8. Ratio of spoil production to coal output in England and Wales 1920–1980 (Whitbread et al. *1991)*

Year	Coal output* (million tonnes)	Spoil output†	Spoil output/ coal output (%)
1920	230	9	4
1930	240	12	5
1940	220	18	8
1950	200	15	8
1960	185	38	21
1970	135	55	41
1980	105	50	48
1988/89	85	45	53

* Does not include output from open-cast mining
† Does not include spoil disposed of in the sea

Fig. 12. Colliery spoil tip

The most common minerals in colliery spoil are quartz, mica and clay minerals and lesser quantities of pyrites and carbonates of calcium, magnesium and iron. Oxidation of the pyrites, which is one of the causes of spontaneous combustion in colliery spoil (see below), can mean that some of the carbonates are converted into their corresponding sulphates. Typical chemical analyses of burnt (Sherwood and Ryley 1970) and unburnt spoil (Rainbow 1989) are given in Table 9. The physical and chemical properties of a range of spoils are given in Table 10 and in Figs 13–19.

Table 9. Chemical composition of colliery spoil (Sherwood and Ryley 1970; Rainbow 1989)

Component	Burnt spoil (%)	Unburnt spoil (%)
SiO_2	45−60	37−55
Al_2O_3	21−31	17−23
Fe_2O_3	4−13	4−11
CaO	0·5−6	0·4−4·9
MgO	1−3	0·9−3·2
Na_2O	0·2−0·6	0·2−0·8
K_2O	2−3·5	1·6−3·6
SO_3	0·1−5	0·5−2·5
Loss on ignition	2−6	10−40

Table 10. Physical and chemical properties of some colliery spoils (Sherwood and Ryley 1970, and unpublished results of British Coal and TRL)

Particle diameter (mm)	Burnt spoils						Unburnt spoils				
	A	B	C	D	E	F	S	T	U	V	W
				Particle size distribution (%)							
>40	2	0	6	0	3	0	7	5	6	12	15
20−40	22	14	14	20	18	15	33	25	7	18	15
10−20	26	22	23	23	22	25	26	35	10	15	32
5−10	22	21	22	20	19	17	17	16	15	15	18
2−5	11	12	13	12	12	11	6	6	17	10	8
<2	17	31	22	25	32	32	11	13	45	30	12
				Particle density (Mg/m^3)							
	2·65	2·69	2·71	2·72	2·76	2·90	2·60	2·51	— Not determined —		
				pH of shale-water suspension							
	6·5	6·8	5·4	4·2	4·5	8·5	— Not determined —				
				Soluble sulphate (g.SO$_3$/litre)							
	0·6	1·4	1·6	7·0	6·9	1·5	— Not determined —				

Chemical problems

Spontaneous combustion

Spontaneous combustion is a hypothetical rather than a real problem, but it is a good example of a problem that is unique to waste materials as it could never arise with the naturally-occurring materials that are normally used in road construction. The problem originates from the presence of coal in unburnt colliery spoil and the possibility of this being ignited by exothermic reactions

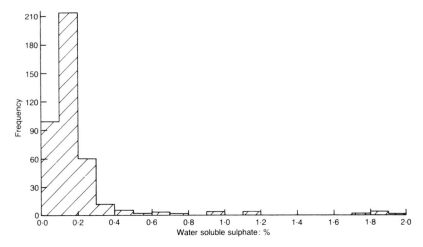

Fig. 13. Range of water-soluble sulphate content values of unburnt colliery spoils (Rainbow 1989)

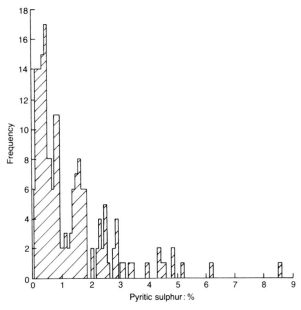

Fig. 14. Range of pyritic sulphur content values of unburnt colliery spoils (Rainbow 1989)

in the spoil, such as the oxidation of pyrites. In the past there has been concern about the use of this material because of the suspected risk, based on observation of burning and burnt-out spoil heaps. This is reflected in the Specification for Highway Works (1991a) which has a requirement that bulk fill material susceptible to spontaneous combustion should not be used. However, experience over some 25 years has shown that there is no risk because even if combustible

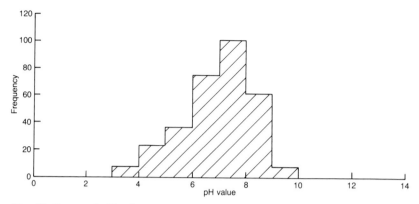

Fig. 15. Range of pH values in unburnt colliery spoils (Rainbow 1989)

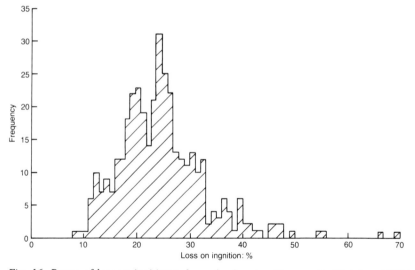

Fig. 16. Range of loss on ignition values of unburnt colliery spoils (Rainbow 1989)

material is present the required state of compaction within an embankment is such that the air content is too low to allow combustion. This fact is recognized by the Department of Transport which does allow unburnt colliery spoil to be used for bulk fill.

Sulphates

The soluble sulphate content of unburnt spoils is generally too low to be a serious problem (see Fig. 13) but soluble sulphates may occur in large concentrations in burnt colliery spoil due to the oxidation of pyrites in the unburnt spoil during combustion (Sherwood and Ryley 1970). Sulphates may cause problems by migrating from the spoil and reacting with the cement in

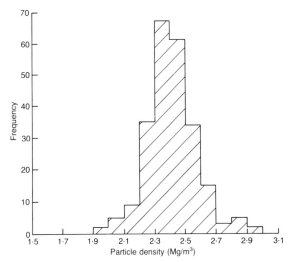

Fig. 17. Range of particle density values of unburnt colliery spoils (Rainbow 1989)

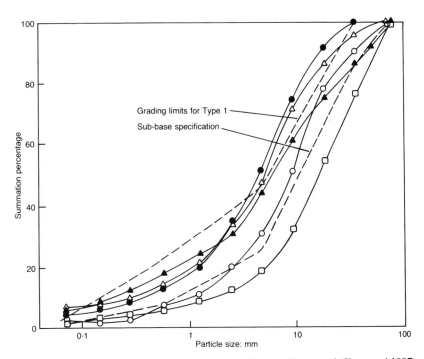

Fig. 18. Particle size distribution of some samples of burnt colliery spoil (Sherwood 1987)

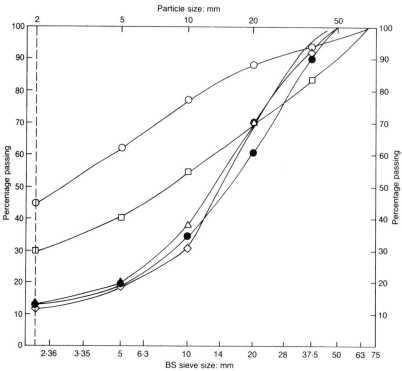

Fig. 19. Particle size distribution of some samples of unburnt colliery spoil (from British Coal)

concrete and other cement-bound materials to form products that occupy a greater volume than the combined volume of the reactants. With all forms of sulphate attack, water is an essential part of the reaction. The water present in the spoil at the time of placing will be insufficient to dissolve very much sulphate, so unless extra water is able to enter the material no appreciable attack will occur even if high concentrations of sulphate are present.

The expansive reaction between sulphates and concrete has been known for at least 100 years and the problem is well understood. It arises because of the ability of sulphates to react, in the presence of water, with the hydrated cement in the concrete to form calcium sulphoaluminate (ettringite). This occupies a greater volume than the combined volume of the reactants, which leads to the expansion and disintegration of the concrete. The main reaction is between calcium sulphate ($CaSO_4$, $2H_2O$) and the hydrated calcium aluminates:

$$3CaO \cdot Al_2O_3 \cdot 19H_2O + 3(CaSO_4 \cdot 2H_2O) + 6H_2O$$
$$= 3CaO \cdot Al_2O_3 \cdot 3CaSO_4 \cdot 31H_2O$$

Magnesium and sodium sulphates behave in a similar manner but are more harmful because of their higher solubilities. Magnesium sulphate is also able to react with the hydrated calcium silicates.

The reactions between sulphates and the hydrated silicates and aluminates lead to products that occupy a greater volume than the combined volume of the reacting constituents. Although the role of these reactions in initiating the expansion is not in question, Lea (1970) pointed out that there are difficulties in ascribing the observed expansion directly to the increased volume of solids. This is because the expansions that occur are much greater than would be expected from this cause alone. He reviewed the possible reasons that have been put forward for this discrepancy, which include osmotic effects and secondary expansion resulting from the destruction of the cementitious materials.

Determination of sulphates. Determination of the total sulphate content of colliery spoil is relatively easy. The method given in BS 1377: Part 3 (1990) for soils may also be used for colliery spoil; it involves extraction of the sulphates in the spoil with dilute acid followed by gravimetric determination as barium sulphate. A similar method, for the determination of the total sulphate content of aggregates, is included in BS 812: Part 118 (1988). This latter standard includes data on the precision, which is quite good, of the test method.

Sulphate attack of concrete can only occur if the sulphates are able to migrate from the spoil and attack concrete in the vicinity of the placed spoil. Determination of the total sulphate content of the spoil is thus of limited value because it gives no indication of the potential for the sulphate to pass into solution. Measurement of the water-soluble sulphate content is therefore the preferred method for determining the degree of risk which sulphates may present.

The usual procedure is to extract the sulphate ions from the spoil with a limited amount of water to restrict the influence of the sparingly soluble calcium sulphate on the result. BS 812: Part 118 (1988), BS 1047 (1983a) and BS 1377: Part 3 (1990) all give basically the same method which involves the removal of sulphate ions by shaking one part by mass of the material with two parts by mass of water and expressing the sulphate content (as SO_3) of the aqueous extract in terms of grams per litre (g.SO_3/litre) (note: the soluble sulphate content is occasionally reported as a percentage, as in Fig. 13 — 2g.SO_3/litre is equivalent to $0 \cdot 2\%$ of SO_3). The solubility of calcium sulphate is only $1 \cdot 2$ g.SO_3/litre and, if the sulphate ion concentration exceeds this value, other more soluble sulphates must be present. The Specification for Highway Works (1991a) requires that all, fill and unbound sub-base and base materials, 'when placed within 500 mm of cement-bound materials, concrete pavements and concrete products they shall have a soluble sulphate content not exceeding $1 \cdot 9$ g. of sulphate (expressed as SO_3) per litre'.

Sulphides

Sulphides in the form of iron pyrites (FeS_2) frequently occur in unburnt colliery spoil (see Fig. 14) but are less likely to be present in burnt spoil because they become oxidized to sulphates during combustion. Iron pyrites is of little concern but when exposed to air it may oxidize to jarosite ($KFe_3(SO_4)_2(OH)_6$) by the following sequence of reactions:

$$2FeS_2 + 2H_2O + 7O_2 = 2FeSO_4 + 2H_2SO_4 \tag{1}$$

$$4FeSO_4 + O_2 + 2H_2SO_4 = 2Fe(SO_4)_3 + 2H_2O \qquad (2)$$
$$3Fe_2(SO_4)_3 + 12H_2O = 2HFe_3(SO_4)_2(OH)_6 + 5H_2SO_4 \qquad (3)$$

Base exchange reaction with potassium minerals in the spoil then leads to the formation of jarosite. If calcium carbonate is present the sulphuric acid released by (1) and (3) will react to form gypsum:

$$CaCO_3 + H_2SO_4 = CaSO_4 \cdot 2H_2O + CO_2 \qquad (4)$$

Most of these reactions are exothermic and the heat produced is one of the reasons for spontaneous combustion in unburnt spoil. As the reactions produce sulphates it follows that they may then attack concrete and other cementitious materials in the manner considered above. However, apart from this, additional expansive reactions may arise which could cause heave in an exposed formation even if no concrete is present. The calculated volume expansion from pyrites to jarosite is reported to be 115%, and the volume expansion from pyrites to ferric sulphate is 170%. The formation of gypsum is accompanied by a large increase in volume and this is believed to be the main cause of expansion in pyritic shales, which has occasionally proved to be a problem when such shales have been used in construction (Nixon 1978, Hawkins and Pinches 1987). Instances have been reported where heave occurred even though little calcium carbonate was present (Collins 1990).

Determination of sulphides. A method for the determination of the total sulphur content contributed by sulphates and sulphides is included in BS 1047 (1983a) and the method could also be applied to colliery spoil. In this method the sulphides are oxidized to sulphates with a suitable oxidizing agent and the combined sulphate content is determined by the method described above for sulphates. The value obtained, which is expressed in terms of the total sulphur content, is made up of the initial sulphate content plus the contribution to the sulphate content resulting from the oxidation of the sulphides. The sulphide content of the material may then be readily calculated if a separate determination is also made of the original sulphate content of the material before the oxidation of sulphide.

Uses of colliery spoil in road construction
Bulk fill
Colliery spoil has been successfully used as bulk fill material. At the peak of the motorway building programme in the early 1970s it was estimated that about 8 million tonnes per year were being used. This was the largest amount of any waste material or industrial by-product being used in road construction and it represented the biggest single useful commercial outlet for colliery spoil.

The main problem with colliery spoil as a fill material is its variability within a deposit. Spoil heaps may contain burnt spoil, partially-burnt spoil, unburnt spoil and mine tailings, with quantities of all four occurring quite close together. However, except for mine tailings, all are suitable for fill. Visual inspection of the tip should ensure that the type of material delivered to the site does not vary too frequently since control of compaction may be difficult. When considering compaction requirements on the basis of the recommendations given in the Specification for Highway Works (1991a), most unburnt spoils are classified as 'well-graded granular and dry cohesive soils'. However, some, while being acceptable for use as fill, may have untypically high fines or moisture contents and could more appropriately be considered as 'cohesive soils'.

Sulphates. Normally the presence of sulphates in bulk fill will not be a problem; it becomes a problem only if the fill is placed close to concrete structures when the limits given above apply.

Frost susceptibility. Most unburnt spoils are not susceptible to frost when tested by BS 812: Part 124 (1989). On the other hand, burnt spoils are usually highly susceptible to frost. However, frost susceptibility of the compacted fill is unlikely to be a problem if the fill is more than 450 mm below the finished road surface. In the UK frost penetration rarely exceeds 450 mm, and in all but the most lightly-trafficked roads the fill material will be below this depth. If the fill material is likely to be subjected to frost penetration a check should be made to ensure that it is non-susceptible to frost-heave as defined by the BS test.

As selected granular fill

Table 4 showed that unburnt spoil is excluded from use as a selected granular fill material. This is not surprising as, although it can on occasion seem to be granular, it does not possess the properties generally associated with such materials. The coarse particles are not discrete and are usually aggregations of smaller particles, which means that the long-term stability of the aggregated particles is open to question.

Well-burnt spoil is, however, specified by name for many applications of selected granular fill. Fig. 18 shows that burnt spoil can be obtained which meets the grading requirements for granular sub-base material and can therefore easily satisfy the more relaxed grading requirements of selected granular fills.

Burnt spoil is excluded from some uses but the exclusions can be justified. More surprising than the exclusions are the inclusions where, for example, burnt colliery spoil is permitted to be used as fill to Reinforced Earth (6I), the upper bed (Class 6L) and surround (Class M) to steel structures. There is no danger of unsuitable spoils being used as the specification also imposes limits for pH, sulphate and chloride content and redox potential. However, these limits are such as to make it virtually impossible for most spoils to meet the specification.

As selected cohesive fill

Table 4 showed that the Specification for Highway Works (1991a) excludes unburnt spoil from use as a selected cohesive fill for all categories except as fill to structures (Class 7A). The omission of unburnt spoil from the materials that are potentially suitable for stabilization with lime (Class 7E) or with cement (Class 7D) is open to question and will be considered further.

The exclusion of unburnt spoil as a fill material to Reinforced Earth (Class 7C) has been critically reviewed by West and O'Reilly (1986). They pointed out that the failure of a Reinforced Earth structure could have serious consequences and it is therefore essential that all precautions should be taken to ensure that the structure will perform satisfactorily over its design life. As there is a small, but significant, risk that this would not be the case if unburnt spoil were to be used as the fill for Reinforced Earth, they endorsed its exclusion from the list of suitable materials. They added that in any case the potential outlet for unburnt spoil as a fill material for Reinforced Earth is infinitesimal in relation to the outlet for its use in bulk fill.

As a granular capping material

The requirements for unbound granular capping materials (6F1 and 6F2) as given in the Specification for Highway Works (1991a) were listed in Table 4. Unburnt spoil is logically excluded but burnt spoil is permitted for use provided that it meets the requirements set out in Table 4. No limits for the sulphate content are given but an overriding factor would be the general restriction on the maximum sulphate content of any material placed in proximity to concrete. If the capping were to come within 450 mm of the road surface it would also be necessary to check that the material was not frost-susceptible.

As a stabilized capping material

Burnt spoil could be stabilized with cement to form a stabilized capping material, but generally speaking there would be little reason for doing so as it would be suitable for use in unbound form.

Unburnt spoil is excluded by the Specification for Highway Works (1991a) from use as a stabilized material for capping layers. This can be justified but its exclusion is, at first sight, anomalous because cement-stabilized spoil is permitted to be used for sub-base construction provided that it fulfils all the requirements of clause 1036 for cement-bound material category 1 (CBM1). It thus does not seem logical to permit it for sub-base while prohibiting it from less onerous use in a capping layer.

Part of the reason lies in the manner in which the clauses for cement-stabilized capping (clause 614) and CBM1 (clause 1036) are drafted. Of the two the latter is much preferable because it specifies the properties of the material and the stabilized end-product in considerable detail but does not define the actual material to be used. Hence, there are no restrictions on the material as such but requirements on its properties before and after it has been stabilized. Clause 614 on the other hand, attempts to define the materials to be used but gives inadequate detail as to the property of the stabilized material that is being sought.

There is much evidence to show that unburnt colliery spoil can be successfully stabilized with cement (Kettle and Williams 1978, Tanfield 1978) and a guide on the subject was published by the National Coal Board (1983). Well-publicized failures, due to expansion of the cement-stabilized material after compaction, have occurred with its use (*New Civil Engineer* 1980) and this may explain why it has been excluded. However, the testing regime which is specified for cement-stabilized sub-base and base materials has, since 1986, included a durability requirement which should detect whether any problems are likely to occur with cement-stabilized materials. It is for this reason that the Specification for Highway Works (1991a) does not attempt to define the actual materials that are to be used for CBM1 sub-base materials.

The ability of the durability requirement, given in the specification, to detect whether problems with the expansion of cement-stabilized spoil were likely to arise has been confirmed by Thomas *et al.* (1987). Carr and Withers (1987) showed that two types of expansion can occur in cement-stabilized spoil: One, in the short-term, is due to hydration of the clay minerals within the spoil, the other, which can occur in the longer-term, is due to sulphate attack of the cement matrix.

It is probable that unburnt spoil could be stabilized with lime because it contains clay minerals that can react with lime. However, as little work has been done on this subject, lime-stabilized unburnt spoil is excluded as a capping material.

As an unbound sub-base material

'Well-burnt non-plastic shale' is one of the materials mentioned by name in the Specification for Highway Works (1991a) as being acceptable, as a granular sub-base material. The problem lies with the definition of 'well-burnt' as there are large differences in the physical properties of well-burnt and unburnt colliery spoil which can make the former eminently suitable for use as a granular sub-base and the latter totally unsuitable. The main difficulty in using spoil as an unbound granular sub-base material, therefore, lies in distinguishing how well burnt it is and in avoiding materials that are only partially burnt. Colour is some indication but is not wholly reliable. Some red colliery spoils which appear well-burnt may have only been fired on the outside and may have 'black hearts'. Conversely, some well-burnt spoils may have a black appearance due to their having been fired in a reducing atmosphere, but this does not necessarily imply unsuitability. In cases of doubt, hitting the aggregate with a hammer can distinguish well-burnt from unburnt materials as the former ring when struck compared to the dull sound emitted from unburnt spoil.

Figure 18 shows that burnt colliery spoil may be obtained with a particle size that satisfies the grading requirements for granular sub-bases. Unfortunately, however, recent changes to the Specification for Highway Works (1991a) make it doubtful if it could meet the strength and durability requirements that are now specified. All unbound granular sub-base materials are now required to have a minimum soaked 10% fines value (TFV) of 50 kN. It is doubtful if many spoils could meet this requirement, even though in practice they would be suitable. Up to the publication of the 1986 edition of the specification this was recognized and well-burnt shale was exempted from this requirement.

The imposition of a requirement for a minimum TFV originates from work by Hosking and Tubey (1969). This work was carried out on natural aggregates most of which pass the TFV requirement with ease. Dawson and Bullen (1991) have shown that the requirement unjustly excludes furnace bottom ash for use as an unbound sub-base material (see chapter 5) and the same is true of burnt colliery spoil.

The latest edition of the Specification for Highway Works (1991a) has made

Table. 11. Effect of the addition of cement on the frost-heave of burnt colliery spoil (Fraser and Lake 1967)

Sample no.	Frost-heave (mm)	
	Without cement	With 5% cement
1	22	7
2	41	13
3	45	11
4	30	8
5	6	0

it still more difficult for burnt colliery spoil to fulfil the requirements for unbound sub-base materials as it now includes a durability requirement in terms of the magnesium soundness value (see Fig. 9). This is a result of research by Bullas and West (1991) aimed at defining the terms 'clean, hard and durable' which are often used as a subjective description of aggregates to be used in road construction. For aggregates used in bitumen macadam road base, Bullas and West (1991) suggested a magnesium sulphate soundness value of 75, as determined by the method given in BS 812: Part 121, as the criterion for durability. Although this research was specifically aimed at bitumen macadam road base and did not deal with the durability of unbound sub-base materials, the 1991 edition of the specification stipulates the same sulphate soundness value of 75 for unbound granular materials.

The reasons for imposing more stringent requirements on burnt spoil are not given, and for minor roadworks there would seem to be no reason why the specification in use up to 1986 should not continue to be used. This would dispense with the need to determine TFV and soundness.

Burnt colliery spoil frequently contains sulphates in sufficient quantity to preclude its use as a sub-base in situations where it is within 500 mm of concrete structures or road pavement layers containing cement. It is also usually highly susceptible to frost, which means that it should not be used within 450 mm of the road surface. However, Table 11 shows that the frost susceptibility may be reduced by the addition of cement.

As a cement-bound sub-base material

Provided they can meet all the relevant requirements, the Specification for Highway Works (1991a) permits the use of both unburnt and burnt colliery spoil as the 'aggregate' in CBM1 and CBM2. As mentioned above this gives rise to a slight anomaly in that cement-stabilized unburnt spoil is *not* permitted to be used as a capping layer.

The presence of high sulphate contents may occasionally cause problems. There is no requirement in the specifications for CBM1 and CBM2 for the sulphate content to be determined as it is assumed that problems arising from the presence of sulphates would be demonstrated by failure of the test specimens to meet the soaking requirement given in Fig. 10.

Other uses

There is very little potential for colliery spoil to be used above sub-base level. Burnt spoil might meet the grading and strength requirements for CBM3 and CBM4 cement-bound road base material given in Fig. 10, but it is specifically excluded from use.

4. China clay wastes

Occurrence

China clay (kaolin) is used in the paper and ceramic industries. Most of the world's production is in South-West England and in North Carolina and Georgia in the USA, Britain being by far the largest producer. In South-West England, the commercial extraction of china clay (kaolin) from kaolinized granite is concentrated in the St Austell area with subsidiary workings in the nearby Bodmin and Lee Moor areas.

The kaolin was formed geologically by the action of steam and carbon dioxide on the orthoclase feldspar as the granite cooled and is extracted from steep-sided open pits by subjecting the face to high-pressure jets of water (see Fig. 20). The broken-up rock flows in a slurry to the pit bottom from where it is pumped to a separating plant. Here the bigger grains, which are predominantly quartz with small but variable amounts of other minerals, including a few flakes of mica, are separated from the sand waste. The residual slurry is dewatered and a second separating process removes the fine clayey sand and mica residue. Thus, for each tonne of china clay produced, about 9 tonnes of waste are also produced. This waste material is composed of approximately:

- 2 tonnes of overburden
- 2 tonnes of waste rock (stent)
- 3·7 tonnes of coarse sand waste
- 0·7 tonnes of micaceous residue.

Current mining regulations and other factors tend to make the extraction of china clay at depths much in excess of 45−60 m uneconomic, although the kaolinization continues to greater depths. Deeper workings are technically feasible and to avoid sterilizing these reserves from possible future reworking the pits are not normally back-filled. Further, the production of kaolins to special requirements often needs the blending of kaolins from different sources and this is facilitated if the pits remain available for sporadic reworking as required.

Where suitable, the pits are used for storing water, which is required in large quantities for extracting the china clay. The wastes are tipped, usually on land less suitable for china clay working.

In the past all the waste residues were deposited on the same tips, which is unfortunate because the coarse sand waste has much more potential for use

Fig. 20. China clay workings

than do the other materials. More recently some of the granular material has been tipped separately, making it easier to extract.

Gutt *et al.* (1974) estimated that the total stockpile of wastes on the tips was 280 million tonnes. A more recent estimate gives a figure of 350 million tonnes for the stockpile (Whitbread *et al.* 1991). The stockpile continues to grow as the current production of wastes is estimated (Whitbread *et al.* 1991) to be 27 million tonnes per year and very little is used. The china clay industry is thus second only to British Coal as an originator of industrial wastes in the UK and it has the third largest stockpile; only the stockpiles of colliery spoil (3600 million tonnes) and of the almost defunct slate industry (400 million tonnes) are comparable in quantity.

Additionally it is the most concentrated stockpiling of the industrial waste materials and a report commissioned by the Department of the Environment (1991b) concluded that: 'The china clay industry has particular problems with its very high level of permanent waste which it cannot return to the excavation. The tips are unsightly to many people . . . and represent major issues which have yet to be resolved'.

Although 30−40% of Cornwall County Council's aggregate requirements are met by using china clay wastes, the demand for the wastes is insignificant compared to the annual output. As china clay production continues the stockpile is expected to increase at a substantial rate (Hocking 1994).

Composition of china clay wastes

Sand

Of the wastes produced by the extraction of china clay, the coarse sand is not only the largest component, but also that with the most desirable engineering properties. The sand is largely composed of quartz (85−88%), with smaller proportions of tourmaline (9−11%), feldspar (2−3%) and mica (1−2·5%). It is a quartzitic sand which is chemically inert; even if it did contain any soluble salts, most would be removed during the extraction process and any that remained would be rapidly leached out of such a free-draining material. A typical chemical analysis is given in Table 12.

The particle size distributions of typical sands are given in Fig. 21. The gradings of sands vary from tip to tip but it is possible to choose sands that meet the specifications for many roadmaking purposes.

Table 12. Chemical composition of china clay sand (Hocking 1994, OECD 1977)

Component	Composition (%)
SiO_2	75−90
Al_2O_3	5−15
Fe_2O_3	0·5−1·2
TiO_2	0·05−0·15
CaO	0·05−0·5
K_2O	1−7·5
Na_2O	0·02−0·75
MgO	0·05−0·5
Loss on ignition	1·2

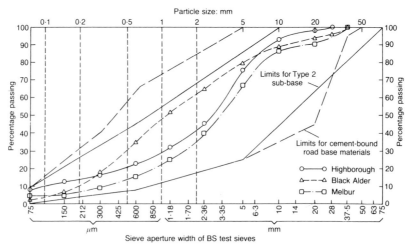

Fig. 21. Particle size distribution of samples of china clay sand (Tubey 1978)

Stent

The stent, which can vary in size from < 100 mm to in excess of 2 m in diameter, essentially consists of massive quartz, quartz/tourmaline and partially kaolinized granite. The irregular distribution of these materials within the rock mass inevitably gives rise to variability in the waste. Typical properties reported by Hocking (1994) are:

- saturated TFV 75−150 kN
- magnesium sulphate soundness value 70−90%
- soluble sulphate content < 0·1%

Uses of china clay sand in road construction

Due to its abundance and superior properties, china clay sand is the most important of the waste materials of china clay production. The sand is a good quality aggregate which needs only the same basic grading and washing processes that are applied to other natural aggregates. it has accumulated in stockpiles only because of the large amounts produced in an area where there is low demand.

Bulk fill

At first sight china clay sand would appear to be an excellent material for bulk fill, but there is a reluctance to use it in road construction. The reason for this is that micaceous materials have a dubious reputation among road engineers and the literature contains reports of compaction difficulties associated with their use. The obvious presence of mica in china-clay sand has, therefore, given rise to suspicions that similar difficulties may occur when they are used in road works. However, compaction tests on selected sands (Tubey 1978) suggested that the mica present was not having any serious deleterious effects on the compaction.

Changes in density can be caused by the presence of mica. The larger change is produced solely as a result of the change in grading produced by the presence of mica. The other change is due to some physical property which is unique to mica; this is almost certainly the resilience of the mica flakes which allows them to deform under load and to recover after the load has been removed, rather like the leaves of a leaf spring.

Research by Tubey and Webster (1978) on the effect of mica on the compaction properties of a range of materials showed that the resilience of the mica reduced the state of compaction achievable for a given compactive effort by about $0·007$ Mg/m^3 for each one per cent of fine and coarse mica, respectively.

The distinctive colour, lustre and thin flaky shape of mica makes its presence, even in trace amounts, very obvious. It is possible, therefore, that many of the difficulties attributed to its presence in the past may have been caused by other factors, such as overall particle size distribution.

Despite the alleged problems the sand has been used with considerable success as bulk fill for earthworks where a strict moisture content control was imposed to enable adequate compaction to be achieved (Hocking 1994). When dry the material requires wetting, and when excessively wet the residual clay and mica content give the fill an apparent thixotropy which eventually disperses when drained. Once compacted the material develops a natural 'set' due to the cementing action of the clay/mica, resulting in a marked increase in stability.

As selected granular fill

Table 4 showed that sands and gravels are acceptable constituents of Class 6 granular fills. China clay sand is therefore potentially suitable for such use and would almost invariably satisfy the requirements for plasticity and chemical composition. Whether or not it satisfied the grading and particle strength requirements would depend on particular circumstances. Fig. 21 shows that by careful selection china clay sand can be obtained which will satisfy the more stringent requirements for unbound granular sub-bases, and it has been used for this purpose. It therefore follows that it should not be too difficult to find sources of china clay sand which will meet the requirements for selected granular fills. However, selection of suitable materials from a stockpile inevitably increases the price.

As a granular capping material

The requirements for unbound granular capping materials (6F1 and 6F2) as given in the Specification for Highway Works (1991a) were summarized in Table 4. This showed that china clay sand is potentially suitable for use in granular capping layers. As it has been used successfully as an unbound granular sub-base material in selected applications, it clearly has a higher potential for use as a granular capping.

As a cement-stabilized capping material

If it does not meet all the grading requirements for an unbound granular capping, china clay sand can be upgraded by stabilization with cement. No problems of excessive plasticity or with chemical composition are likely to arise with china clay sand and it mixes easily with cement. The results in Table 13 show that it can be stabilized with cement to meet the strength requirements for cement-bound sub-bases.

As a granular sub-base material

The Specification for Highway Works (1991a) excludes natural sands and gravels from use as Type 1 granular sub-base materials but they are permitted for use as Type 2 sub-base. Fig. 21 shows that the coarser varieties of china clay sand meet the grading requirements, given in Fig. 9, for Type 2 materials. The sand is not likely to present any problems with regard to chemical composition, particle strength and durability. Tubey (1978) showed that china clay was marginally frost-susceptible, but the criteria for frost-susceptibility have since been relaxed and frost-susceptibility is unlikely to be a problem.

As a cement-bound sub-base and base material

The results in Fig. 21 and Table 13 show that china clay sand can be stabilized with cement to produce a material that complies with the requirements for CBM1 and CBM2 cement-bound sub-base material with regard to both grading and strength. The sand would also be potentially suitable for use as the fine

Table 13. Compressive strength of cement-stabilized china clay sands (Tubey 1978)

Source of sand	Cement content (%)	7-day strength (MPa)
Melbur	5	2·8
Melbur	10	7·4
Black Alder	5	4·4
Black Alder	10	13·4
Highborough	5	4·8
Highborough	10	11·8

aggregate in CBM3 and CBM4 cement-bound road base but might require processing to produce material of the correct grading.

As a concreting sand

China clay sands are used in South-West England as mortar and concreting sands and also for the manufacture of concrete blocks. However, when used in concrete they have a high water demand leading to a reduction in strength which has to be compensated for by increasing the cement content. The concretes are also said to be harsh in texture and to have a poor workability which makes them more difficult to place (BACMI 1991).

Use of stent in road construction

The use of crushed stent as a substitute for primary aggregates in road construction has been encouraged in recent years in Cornwall (Hocking 1994). Using conventional crushing and screening plants it has been possible to produce aggregates that meet the majority of requirements for aggregates given in the Specification for Highway Works (1991a). These include drainage filter media, pipe bedding, selected granular fill and Type 1 sub-base. It has also been used to produce coarse aggregates for concrete in accordance with BS 882 (1992).

5. Power station wastes (pulverized fuel ash and furnace bottom ash)

Occurrence

About 10 million tonnes per year of pulverized fuel ash (PFA) and $2 \cdot 5$ million tonnes per year of furnace bottom ash (FBA) are currently produced in the UK. The output and utilization of PFA and FBA in 1989−90 is given in Table 14. The amount produced is expected to decline with the run-down of coal-fired power stations and the increasing proportion of electricity produced by gas-fired and nuclear power stations.

The policy of the electricity generating companies has been to site their power stations as close to the coalfields as is practicable. This means that most of the larger modern power stations are in central England, and Fig. 22 shows that this area produces more than half of the available supplies.

Coal-burning power stations use coal which has been pulverized to a fine powder. When the pulverized coal is burned in a furnace at the power station it produces a very fine ash which is carried out of the furnace with the flue-gases. This ash is PFA (or fly-ash) and accounts for about 75−85% of the ash formed from the burnt coal. The remaining, coarser fraction of the ash falls to the bottom of the furnace where it sinters to form the FBA.

PFA is removed from the flue-gases by mechanical and electrostatic precipitation and is initially collected in hoppers; it can be supplied as dry powder (hopper ash) in this state. When required this ash can be passed through a mixer−conveyor plant where a measured amount of water can be added. It is then known as 'conditioned PFA' which can be stockpiled if it is not required immediately. At certain power stations PFA is transported

Table 14. Ash output and utilization in the UK 1989−90 (Whitbread et al. 1991)

	Ash production (million tonnes)			Ash sales (million tonnes)	Percentage of production
	PFA	FBA	Total		
National Power	6·0	1·5	7·5	3·4	45·3
PowerGen	3·8	1·0	4·8	2·2	45·8
Scottish Power	0·7	0·1	0·8	0·1	13·5
Total	10·5	2·6	13·1	5·7	43·5

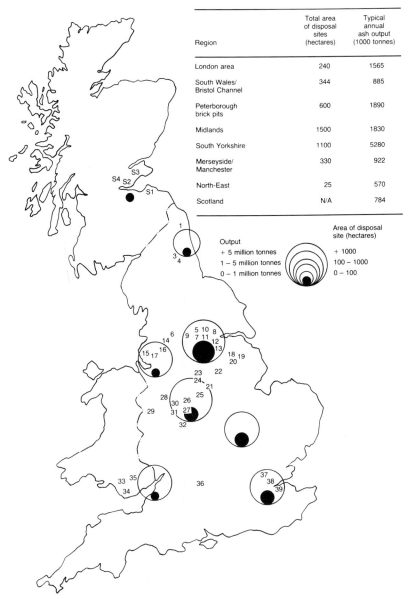

Region	Total area of disposal sites (hectares)	Typical annual ash output (1000 tonnes)
London area	240	1565
South Wales/ Bristol Channel	344	885
Peterborough brick pits	600	1890
Midlands	1500	1830
South Yorkshire	1100	5280
Merseyside/ Manchester	330	922
North-East	25	570
Scotland	N/A	784

Fig. 22. Distribution of coal-burning power stations and sources of PFA (Department of the Environment 1991)

hydraulically to lagoons from where it can be later reclaimed; it is then known as 'lagoon PFA'.

At some power stations FBA is washed out to lagoon storage with PFA. Due to separation during sedimentation in the lagoons, lagoon ash is more variable in composition and is very much coarser.

Composition of PFA

PFA consists of glassy spheres together with some crystalline matter and a varying amount of carbon. The overall colour ranges from almost cream to dark grey and is affected by the proportions of carbon, iron and moisture. The three predominant elements in PFA produced by burning British coals are silicon, aluminium and iron, the oxides of which together account for 75–95% of the material. Such ashes are known as alumino-silicate fly-ash. There is also another variety, known as sulpho-calcitic fly-ash, which is produced by the combustion of coal with a high limestone and sulphur content. Sulpho-calcitic ashes have a high lime (CaO) content; they can therefore have hydraulic properties because of the pozzolanic reaction between the components of the ash. A draft European Standard (CEN 1993) for sulpho-calcitic ash for use as a hydraulic binder specifies that the total calcium oxide content shall be between 37% and 58% with the free lime between 18% and 31%.

Typical chemical analyses of British (alumino-silicate) ashes are given in Table 15. Mineralogical analyses of these ashes show that the silicon is present partly in the crystalline form of quartz (SiO_2) and in association with the aluminium as mullite ($3Al_2O_3 \cdot 2SiO_2$), the rest being in a non-crystalline glassy phase. The iron appears partly as the oxides magnetite (Fe_3O_4) and haematite (Fe_2O_3), the rest is in a glassy phase. The greater proportion of PFA is in fact a glass with the glass content varying between 65% and 90%.

Physically, PFA is a fine powder which bears a close resemblance to Portland cement in general fineness and usually also in colour. The physical properties of some samples of hopper ash are given in Table 16. Fig. 23 gives the particle size distributions of these ashes which shows that they are predominantly silt size. For comparison the particle size distributions of two lagoon ashes are also included in Fig. 23, showing the generally much coarser gradings of these ashes.

PFA presents few problems of a chemical nature when used in road construction. It invariably contains sulphates, occasionally in high

Table 15. Chemical analysis of PFA (National Power 1990)

Component	Maximum (%)	Minimum (%)	Typical (%)
SiO_2	52	48	50
Al_2O_3	32	24	28
Fe_2O_3	15	7	10·5
CaO	5·3	1·8	2·3
MgO	2·1	1·2	1·6
Na_2O	1·8	0·9	1·2
K_2O	4·5	2·3	3·6
TiO_2	1·1	0·9	1·0
SO_3	1·3	0·5	0·7
Cl	0·15	0·05	0·08
Loss on ignition	>10	0·5	2·0

Table 16. Chemical and physical properties of some samples of hopper ash (Sherwood and Ryley 1966)

Power station	Specific surface (cm²/g)	Particle density (Mg/m³)	Ignition loss (%)
Castle Donington	1900	1·90	2·3
Cliff Quay	4500	2·17	2·9
Ferrybridge	4700	2·26	2·0
Hams Hall	5000	1·98	1·6
High Marnham	3000	2·15	1·3
Huncoat	3600	2·37	3·2
Stella South	3500	2·11	3·5
Skelton Grange	3600	2·13	1·4

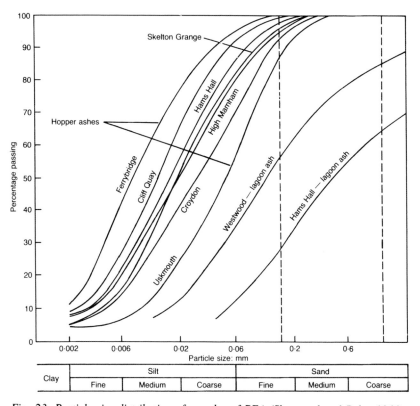

Fig. 23. Particle size distribution of samples of PFA (Sherwood and Ryley 1966)

concentrations. However, because of the low permeability of compacted PFA the presence of sulphates has not been found to be a problem in practice.

Most fresh ashes are highly alkaline, due mainly to the presence of free lime. Carbonation reduces the alkalinity with time but the ash remains well on the alkaline side of neutrality. This does not cause any problems except that metals such as aluminium which corrode in alkaline environments should not be allowed to come into contact with PFA.

Specifications for PFA

As PFA finds considerable uses in cement and concrete, specifications have been prepared giving the requirements for its use in such circumstances. BS 3892 (1983b and 1984) is published in two parts: Part 2 gives the less demanding requirements for PFA. Part 2 divides ashes into grades A and B; grade A is suitable for miscellaneous uses when mixed with Portland cement in concrete where the water requirement and the pozzolanic activity are not normally as great as that obtainable with ash complying with Part 1. Grade B is for use, for example, in precast concrete such as kerbs and flags. The main requirements of the ashes specified in BS 3892 are summarized in Table 17.

Specifications also exist for the properties of PFA used in cement (BS 6588, 1985b and BS 6610, 1985c) and for blended Portland PFA cements (ASTM C595-76, 1976, and BS 6588, 1985b).

Composition of FBA

FBA is the coarser agglomerated material recovered from the bottoms of the combustion chambers of power station boilers fired with pulverized fuel. In appearance it ranges from a highly vitrified, glossy and heavy material to a lightweight, open-textured and more friable type. The precise nature of the material depends on the boiler plant and coal type. It may occur, mixed with PFA, in lagoon ash.

Chemically FBA is very similar to PFA, but in its physical properties it differs entirely, being a coarse granular material ranging in particle size from fine sand to coarse gravel (Fig. 24). The grading makes it potentially suitable as a selected granular fill but because the particles have a porous structure they are relatively weak compared to most granular materials used in road construction.

Table 17. Chemical and fineness requirements of PFA (BS 3892, 1983b and 1984)

	BS 3892: Part 1	BS3892: Part 2	
		Grade A	Grade B
Loss on ignition (%)	<7	<7	<12
Magnesium oxide (MgO) (%)	<4	<4	<4
Sulphate (as SO_3) (%)	<2·5	<2·5	<2·5
Fraction coarser than 45 μm in particle size	<12·5	<30	<60

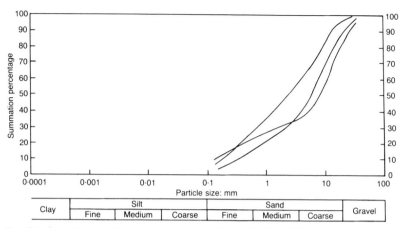

Fig. 24. Particle size distribution of three FBAs (CEGB 1972)

Uses of PFA in road construction
Bulk fill

PFA is a valuable bulk fill material but because of its unusual properties it is given a separate classification (2E) in Class 2 of general cohesive fills listed in the Specification for Highway Works (1991a). Class 2E is specified as reclaimed material from lagoon or stockpile containing not more than 20% of FBA. Certain precautions have to be followed in constructing embankments from PFA and in consequence the specification requires that it shall not be placed within the dimension described in the contract below sub-formation or formation. This is because:

(*a*)   the grain shape and particle size of PFA make the upper layers of PFA difficult to compact

(*b*)   freshly placed PFA behaves in a similar manner to silt and, if not protected, may liquefy under wet conditions

(*c*)   capping and sub-base materials tend to be relatively permeable and a layer of general fill over PFA is considered desirable to add some protection.

The specification also requires the use of a starter layer of Class 6D material below PFA used as bulk fill. This is partly to provide a firm working platform for the construction of the PFA fill. It also functions as a capillary cut-off to inhibit the upward movement of ground water into the PFA. PFA is a silt-sized material with a high capacity for capillary rise, and the possible upward movement of water into the PFA before compaction has occurred can give rise to problems unless the starter layer is included.

Another distinctive feature of the use of PFA as described in the specification is that, unlike other general bulk fill, it is compacted to an end-product specification. Most of the materials considered in this book have particle densities in excess of $2 \cdot 3$ Mg/m^3. Conversely, most ashes produced from coal of British origin have particle densities well below this value. This low particle

Table 18. Maximum dry densities and optimum moisture contents for various fill materials (CEGB 1972)

Type of material	Typical result of a BS compaction test	
	Maximum dry density (Mg/m^3)	Optimum moisture content (%)
Gravel	2·07	9
Sand	1·94	11
Sandy clay	1·84	14
Silty clay	1·67	21
Heavy clay	1·55	28
PFA	1·28	25

density is reflected in the results of compaction tests carried out on PFA which when compacted has a low density compared with most other materials used for mass fill (Table 18). This lightweight property is advantageous when fill material is required on highly compressible soils, and PFA is often specified in these situations.

However, according to Clarke (1992) the particle density of ashes from some British power stations has been increasing in recent years. He claimed that PFA from such stations can no longer be classified as lightweight fill since their particle densities are similar to that of some soils. British coals have not suddenly changed their composition and the claim, if true, is probably due to the increasing use of imported coal.

Many ashes possess self-cementing properties when they are compacted. The result of this hardening, if it occurs, is that settlement within PFA fill is less than with other materials. This makes it particularly useful as a selected fill material behind bridge abutments, where settlement can be particularly troublesome.

PFA can present unusual problems of sampling. For PFA from lagoons the particle size is likely to vary in the lagoon and the size becomes finer with increasing distance from the outfall. To maintain reasonably consistent gradings the material should therefore be excavated in batches at roughly constant distance from the discharge pipes. PFA can also be a problem when conditioned ash is delivered from more than one power station, as the different ashes have widely different properties. For this reason the specification requires that, for each consignment of ash, a record should be kept of the type and source of the material and the name of the power station from which it was obtained.

As a capping material

Although it may have self-hardening properties, PFA used alone is not suitable as a capping layer material. However, as can be seen from Table 19 it may be readily stabilized with cement and the Specification for Highway Works (1991a) lists conditioned PFA as being suitable for this purpose (Class 7G, which, when stabilized with cement, becomes Class 9C). The specification

Table 19. Compressive strengths of cement- and lime-stabilized PFAs (Sherwood and Ryley 1966)

Power station	Compressive strength (MPa)				
	With 10% cement		With 10% lime		
	7 days	28 days	7 days	28 days	56 days
High Marnham	8·35	13·8	1·58	9·95	—
Skelton Grange	5·32	13·1	0·55	5·90	—
Cliff Quay	4·35	8·37	1·03	3·37	—
Ferrybridge	5·62	7·33	3·44	10·8	13·0
Hams Hall	3·38	7·13	0·83	3·79	8·02

requires the PFA to be conditioned ash direct from the power station dust-collection system and to which a controlled quantity of water has been added. Table 19 shows that PFA may also be stabilized with lime but this possibility is not covered by the specification.

As a sub-base material

On its own PFA has no use as a sub-base material but Table 19 shows that, when stabilized with cement, it can be made to meet the requirements of CBM1 cement-bound sub-base material.

As an additive to cement-bound base materials

Dunstan (1981) showed that what he called 'rolled concrete', which was in effect lean concrete (CBM3) with a high addition of PFA, had many advantages over alternative materials for use in dam construction. The properties that were particularly useful were low heat of hydration, ease of compaction, low permeability and good bonding between layers. With the exception of the heat of hydration, which does not create a problem in road construction, the improvement of the properties of lean concrete by the addition of PFA suggested that ash-modified lean concrete might have potential as an alternative to lean concrete (CBM3) for road base construction.

In view of this, the properties of ash-modified lean concrete as a road base material were evaluated (Franklin *et al.* 1982, Harding and Potter 1985, Potter *et al.* 1985). This work showed that the pozzolanic properties (see below for definition) of PFA allowed some of the cement in lean concrete to be replaced with PFA, and although this reduced the early strength the long-term strength was increased. However, the water-reducing properties of PFA could be used to reduce the effect on the early strengths and it was possible to achieve almost any desired strength requirement. This is illustrated by Fig. 25 in which the curve labelled LC is the strength/age relation for normal lean concrete made from a flint—gravel mixture. The curves labelled L1—L5 represent ash-modified lean concretes with the PFA/cement ratios and water/ash + cement ratio (on a volume basis) shown in the accompanying table.

The results in Fig. 25 show that the rate of development of strength of ash-

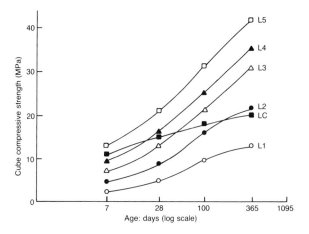

Mix	PFA/cement	Water/ash + cement
L1	4·00	1·62
L2	3·99	1·22
L3	2·76	1·32
L4	1·50	1·60
L5	1·50	1·35
LC	—	3·24

Fig. 25. Strength—age relations of PFA-modified concrete prepared from a flint—gravel aggregate (Franklin et al. 1982)

modified lean concrete is quite different from that of the control. By judicious selection of the mix proportions it is possible to design ash-modified lean concrete that has a strength comparable with that of normal lean concrete at the desired age, but at earlier ages the strength will tend to be lower and the long-term strength much higher. This means that, in principle, ash-modified lean concrete road bases could be designed to have a low early life strength, to produce fine, closely-spaced cracking with good aggregate interlock, and a high in-service strength to ensure good performance under traffic with less likelihood of cracks propagating through the surfacing.

The water-reducing properties of PFA and the spherical shape of its particles also influence the compactibility of ash-modified lean concrete. It is easier to compact than normal lean concrete and the reduction in moisture content results in higher compacted densities being obtained. This means that it can be laid in greater thicknesses than normal lean concrete and satisfactory compaction has been obtained with layers up to 300 mm thick (see Fig. 26).

As an additive to concrete

PFA has been recognized for many years as a valuable material for modifying

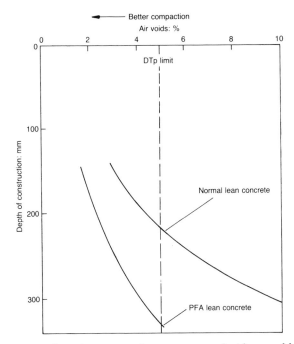

Fig. 26. Density gradients in pavement layers constructed with normal lean concrete and PFA-modified lean concrete (Potter et al. 1985)

and enhancing the properties of concrete. The major benefit to be gained results from the pozzolanic properties of PFA, that is, its reaction with calcium hydroxide produced as one of the hydration products of Portland cement. As this reaction is a consequence of the hydration of cement, it can contribute to the strength of concrete only after hydration has occurred. When ash is used as a partial replacement to cement, the initial strength of the concrete is therefore reduced but the final strength level often exceeds that of a conventional, unmodified mix.

Apart from its pozzolanic action, the addition of PFA to concrete also has significant physical effects. The effects arise partly because PFA contains a large proportion of spherical particles that give a plasticizing action that affects the water demand. In addition, because of the lower particle density of PFA (about 2·4 compared with 3·15 for cement) the introduction of PFA alters the volume of fine material (and hence the volume of cementitious material) by about 10%.

The combined effect of particle shape, grading and particle density causes a substantial reduction in the water demand for concretes containing PFA. In addition to this change, the increased volume of cementitious powder with better packing creates a structure which has the potential for much higher long-term strength, lower permeability and increased resistance to chemical attack. A summary of the effect of PFA on the properties of concrete is given in Table 20.

Table 20. Summary of the effect of PFA on the properties of fresh and hardened concrete (Bamforth 1992)

Property	Effect of PFA
Fresh concrete	
Water demand	Reduced with increasing proportion of PFA
Workability	At a constant water/cement ratio, the slump of PFA concrete is increased. At a constant slump the true workability of PFA concrete, e.g. response to vibration, is increased
Placeability	Based on a qualitative assessment of performance on various construction projects, the addition of PFA improves the placeability
Bleed and settlement	Limited test data showing reduced bleed. No specific reports from site on significant differences between ordinary Portland cement (OPC) and PFA mixes; however, high flow mixes have been used with no bleed problems
Setting time	Increased by the use of PFA, typically by 1 h at 30% level, 2 h at 50%
Finishing	Can be delayed due to extended setting time
Hardening concrete	
Standard cure strength	Early strength gain is reduced but long-term strength is increased
Heat-cycled strength	For equivalent grade mixes, PFA concrete will have a higher heat-cycled strength at 28 days
Air-cured strength	Air-curing affects OPC and PFA concretes to the same extent
Load deformation	Reduced by about 20% at 30% PFA. Creep affected to a greater extent than elastic deformation
Drying shrinkage	Reduced
Durability	
Sulphate resistance	Increased compared with OPC concrete
Acid resistance	Likely to be increased due to lower permeability
Water and gas permeability	Comparable with OPC concrete at 28 days but reduced to a greater extent with age
Carbonation	Unaffected in structures, although short-term laboratory results sometimes indicate an increase
Chloride ingress and salt attack	Chloride ingress is reduced as is deterioration under cyclic salt exposure
Electrical resistivity	Increased substantially

In Portland—pozzolan cements

A pozzolana may be defined (ASTM C595-76 1976) as 'a siliceous, or siliceous and aluminous material, which in itself possesses little or no cementitious value but will, in finely divided form and in the presence of moisture, chemically react with calcium hydroxide at ordinary temperatures to form compounds possessing cementitious properties'. Many naturally-occurring and artificial materials have pozzolanic properties. The natural pozzolanas are for the most part materials of volcanic origin. The artificial pozzolanas are mainly products obtained by the heat treatment of natural materials such as clays, shales and certain siliceous rocks, and PFA.

Calcium hydroxide is one of the hydration reaction products of Portland cement. It contributes nothing to the strength of cemented products and can be a potential source of instability. The addition of a pozzolana to ordinary Portland cement can therefore be beneficial by reacting with the calcium hydroxide released by the hydrating cement to produce further cementitious material. Many of the pozzolanas mentioned above may be employed in the manufacture of Portland—pozzolan cements (ASTM C595-76 1976) but PFA is the most important and specifications exist for both the properties of PFA used in cement, in concrete and for Portland PFA cement.

As alternatives to Portland cement in cement-bound sub-bases and bases

Mixtures of lime and PFA may be used for stabilizing non-cohesive granular materials and, with the most reactive ashes, the lime/PFA mixtures compare favourably in price and performance with Portland cement (Sherwood and Ryley 1966, Gaspar 1976). However, the presence of clay in lime/PFA stabilized soils has a detrimental effect. This is probably because the lime reacts preferentially with the clay fraction to produce products with weaker cementitious properties than the lime PFA reaction products.

Most PFAs contain free lime in the form of CaO. In the presence of water this can react with the siliceous components of the ash so that most ashes have at least some self-hardening properties. Sulpho-calcitic ashes contain sufficient free lime for them to be used as cementitious materials on their own. In southern France an ash is produced which contains up to 25% of free lime. Gravel stabilized with 4% of this ash has been widely used for road construction in the area of the power station (Gaspar 1976). Similar ashes occur in Poland and are used to stabilize gravel for sub-base and base construction.

Uses of FBA in road construction

FBA is a coarse granular material ranging in particle size from fine sand to coarse gravel (Fig. 24). The grading makes it potentially suitable as a selected granular fill and as a granular sub-base material, but because the particles have a porous structure they are relatively weak compared to most granular materials used in road construction. This means that in those instances where a requirement for particle strength in terms of a minimum TFV is specified (see Tables 4 and Fig. 8) it will generally be unable to meet the strength requirements Dawson and Bullen (1991) found that all of the samples of the FBA studied by them had TFVs below 50. However, compaction trials showed that particle degradation of FBA was small and barely changed the amount of fine material in the whole. They therefore concluded that the TFV falsely condemned FBA on the basis of its degradability. This suggests that for minor roadworks the TFV requirement could be abandoned if experience showed that the stability of the material after compaction was satisfactory for the purpose being considered.

6. Blast-furnace slag

Occurrence

Blast-furnace slag is a by-product obtained in the manufacture of pig-iron in a blast furnace, and is formed by the combination of the earthy constituents of the iron ore with the limestone flux. Iron ore is a mixture of oxides of iron, silica and alumina. The chemical reactions in the blast furnace reduce the iron oxides to iron while the silica and alumina compounds combine with the calcium of the fluxing stone (limestone and dolomite) to form the slag. The chemical reactions occur at temperatures between 1300°C and 1600°C produced by the burning of coke which is fed into the furnace along with the ore and the limestone and dolomite. When preheated air is blown into the furnace the oxygen combines with the carbon of the coke to produce heat and carbon monoxide. The iron ore is reduced to iron mainly through the reaction with the iron oxide to produce carbon dioxide and metallic iron. The limestone and dolomite flux is calcined by the heat and dissociates into calcium and magnesium oxides and carbon dioxide. The oxides of calcium and magnesium combine with the silica and alumina of the iron ore to form slag. Thus compounds of lime-silica—alumina and magnesia are formed which collect in a molten stratum above the molten iron in the furnace.

The physical appearance and mineral structure of blast-furnace slag depends largely on the method by which it is cooled. About 70% is air-cooled, which produces a rock-like material with a crystalline structure which when crushed can be used as an excellent substitute for natural aggregates. About 25% is subjected to a water-cooling process where the rapid solidification produces a granulated glassy material. Two other forms exist: foamed slag produced by generating steam within the molten slag from jets of water arranged around the bottom and sides of the pond into which the slag is poured, and pelletized slag, which is similar to granulated slag.

The proportion of slag-to-iron produced by the blast furnace varies with the richness in iron of the ore. In the UK the amount of indigenous ore used for iron production has steadily fallen; by 1987 it had fallen to negligible proportions and all the ore used was imported. The imported ore has a much higher iron content than home-produced ores and this means that even if the production of iron had remained static the amount of slag produced would have fallen. In fact, iron production has also fallen, so less than 4 million tonnes of blast-furnace slag was produced in 1989—90, which is about half that produced ten years earlier. Blast-furnace slag is now produced at only four

Table 21. Blast-furnace slag production (in million tonnes) in the UK 1989–90 (Whitbread et al. 1991)

Source	Air-cooled	Foamed	Granulated	Total
Ravenscraig	0·26	0·13	0·13	0·52
Teesside	0·70	0	0·32	1·02
Scunthorpe	0·80	0	0·30	1·10
Llanwern	0·60	0	0	0·60
Port Talbot	0·40	0	0·20	0·60
Total	2·76	0·13	0·95	3·84

Table 22. Chemical composition of blast-furnace slag (Lee 1974)

Component	Range (% by mass)
CaO	36–43
SiO_2	28–36
Al_2O_3	12–22
MgO	4–11
Total sulphur (as S)	1–2
Total iron ($FeO + Fe_2O_3$)	0·3–2·7

sites (Ravenscraig closed in mid-1992) in the UK; slag production in 1989–90 is given in Table 21.

Composition

The crystalline materials that form blast-furnace slag are compounds of the oxides of calcium and magnesium with silica and alumina. Sulphur is also present as sulphides and sulphates, with minor amounts of iron. The range in the chemical composition of slag is given in Table 22.

Specifications for blast-furnace slag

For most purposes blast-furnace slag is generally regarded as being at least as good as natural aggregates. But, as it differs from these in its chemical and physical properties, the specification and testing methods used for natural materials do not necessarily apply.

BS 812 (1988) is the principal British Standard relating to the testing of aggregates in an unbound form. The physical tests in the standard can, for the most part, be used not only for naturally-occurring aggregates but also for related materials such as slags and crushed concrete. However, the chemical composition of manufactured materials, such as blast-furnace slag, may be markedly different from natural materials, so the chemical tests in BS 812 may not be applicable.

In the case of blast-furnace slag the chemical requirements for its use in construction are given in BS 1047 (1983a). This gives the requirements for the physical and chemical properties of the slag. In the case of the physical properties this is done by reference to the appropriate tests in BS 812. In the case of the chemical properties the standard gives requirements and methods of determining stability (soundness), total sulphur content and water-soluble sulphate content.

Soundness

Two forms of instability can occur. Iron unsoundness is very rare and there is a suspicion that the requirement given in BS 1047 (1983a) is included because it is so easily detected. It arises when partially-reduced iron oxides in the slag oxidize. This is an expansive reaction which causes the slag to disintegrate. It is detected by immersing twelve pieces of slag in water for a period of 14 days and observing whether any of the particles crack or disintegrate.

Calcium disilicate unsoundness, also known as 'falling' and misleadingly as 'lime unsoundness', arises from a phase change when the metastable beta form of calcium disilicate changes to the gamma form. The phase change is accompanied by a 10% increase in volume. Calcium disilicate cannot form in significant amounts when the ratio of CaO and MgO to the SiO_2 and Al_2O_3 and S is kept within the limits given in BS 1047 (1983a). Two conditions given are:

$$CaO + 0 \cdot 8 \, MgO < 1 \cdot 2 \, SiO_2 + 0 \cdot 4 \, Al_2O_3 + 1 \cdot 75 \, S \, (\% \text{ by mass})$$
$$CaO < 0 \cdot 9 \, SiO_2 + 0 \cdot 6 \, Al_2O_3 + 1 \cdot 75 \, S \, (\% \text{ by mass})$$

Slags which satisfy either or both requirements are regarded as sound. Slags which fail to meet these conditions are not necessarily unsound and if the analytical test fails to give a positive answer the decision on whether or not they are unsound is then made solely on the basis of microscopic examination.

Sulphates

Slags used for concrete aggregate have to satisfy the requirements for stability and are also required to have a total sulphur content of not more than 2% (as S) and a sulphate content of not more than $0 \cdot 7\%$ (as SO_3). Slags used in an unbound form have to meet the stability requirements and to have a soluble sulphate content of not more than 2 g/litre.

Other problems

Before the first edition of BS 1047 in 1942 blast-furnace slag had a poor reputation. This problem was effectively dealt with by the new standard so that virtually all new slag production could be adjusted to comply with the standard. The introduction of BS 1047 (1983a) allowed new blast-furnace slags to be used with confidence and in 50 years of utilization there has been no evidence to suggest that this confidence has been misplaced. Moreover, problems with the stability and sulphate content of current blast-furnace slag are rare.

Apart from concern about sulphates and stability, the scope of BS 1047 (1993a) states that it 'excludes requirements for blast-furnace slags from haematite pig-iron (except when dolomite is used as the flux in its manufacture)'. This exclusion has been in the standard for a long time and it is now unclear why it was first introduced. It is in any case highly ambiguous because it can be read in any one of four ways, i.e.

(a)   The slags referred to could not possibly meet the requirements of BS 1047 so that it is a waste of time to test them.

(b)   Whether or not such slags meet the requirements of BS 1047 they are inherently unsuitable and should not be used.

(c)   Such slags may well be satisfactory for use but the tests in BS 1047 cannot be applied to them.

(d)   The slags are known to be satisfactory and do not need to be tested.

In addition to the ambiguous way in which the clause is worded there is a further ambiguity in what is actually meant by 'slags from haematite pig-iron': how can the ore from which the slag was produced be identified? The difference between haematite and other slags, as seen by mineralogy and chemical analysis, is likely to be small compared with the variation within each group. It therefore seems probable that (a) was the reason for the exclusion and if the slag meets the requirements of BS 1047 for stability no problems are likely to arise from this cause provided that it is blast-furnace slag from current production.

However, old deposits of slag are known to give problems which are summarized in BS 6543 (1985a). This states that 'slags from old blast-furnace slag banks are not recommended as fill under buildings unless thoroughly sampled and tested, as they may contain slag wastes other than blast-furnace slag or other industrial wastes. Moreover, there is evidence that some old, partially vitrified slags may react and expand if exposed to sulphate solutions originating either in the ground water or other components of the fill'.

The presence of slags other than blast-furnace slag within a deposit would clearly mean that the requirements of BS 1047 (1993a) (which is specific to blast-furnace slag) would not necessarily apply. This partly explains the warning about old slags given in BS 6435 (1985a), but the standard is not clear as to whether or not a blast-furnace slag which complied with BS 1047 would suffer from the problems associated with sulphates.

Collins (1993) investigated a case of expansion in an old deposit of blast-furnace slag which involved internal sulphate attack. This involved the formation of the mineral thaumasite $[Ca_3Si(OH)_6]_2 \cdot SO_4 \cdot (CO_3)_2 \cdot 24H_2O]$; this mineral is similar to, and isomorphous with, ettringite, the mineral which forms in the sulphate attack of concrete. In this instance the slag did not comply with BS 1047 and Collins concluded that compliance with this standard was a sufficient guarantee of stability. There have been other reported cases of expansion of blast-furnace slag attributable to sulphate reactions but there is no published evidence of whether or not these slags would have been rejected by BS 1047.

Uses of blast-furnace slag in road construction
Air-cooled slag

Some of the physical properties of air-cooled blast-furnace slag are summarized in Table 23.

When the molten slag from the blast furnace solidifies it can be crushed and

Table 23. Physical properties of air-cooled blast-furnace slag (after Lee 1974)

Particle density	$2 \cdot 38 - 2 \cdot 76$ Mg/m^3
Bulk density	$1150 - 1440$ kg/m^3
Water absorption	$1 \cdot 5 - 5\%$ (by mass)
TFV	$70 - 160$ kN
Aggregate impact value	$21 - 42\%$
Aggregate abrasion value	$5 - 31$

screened by normal quarrying procedures to produce a material which satisfies the grading requirements of most specifications. When crushed and screened, the physical properties of the slag make it particularly suitable as an aggregate, both coated and uncoated. It breaks to give consistently good cubical shape and it has a rough surface texture giving good frictional properties and good adhesion to bituminous and cement binders. It is therefore widely used in civil engineering construction as a substitute for naturally-occurring aggregates. This is reflected in the Specification for Highway Works (1991a) which mentions blast-furnace slag by name as being suitable for use in all levels of the road pavement structure, provided that it meets the relevant requirements.

It may be used in an unbound form as selected granular fill and as a granular capping and sub-base material. For example, Table 4 showed that blast-furnace slag can be used as a granular fill for all purposes except where it is used under water or comes into close proximity to metals or as a starting layer below PFA. The use of slag below water could give rise to problems with water pollution and there is a risk of corrosion if slag comes into contact with metals that corrode in an alkaline environment. The exclusion of slag for use as a starter layer below PFA is presumably because of doubts over its drainage characteristics.

There are even fewer restrictions on the use of blast-furnace slag as an unbound granular capping and/or sub-base material, and the self-cementing action of the slag means that, in many ways, it is superior to its naturally-occurring counterparts. Although it can be readily stabilized with cement to meet the requirements of CBM1 and CBM2 sub-base materials, it is seldom necessary to do so.

Blast-furnace slag is permitted for use, with few restrictions, as a CBM3 and CBM4 road base material and as a concrete aggregate. It is also used as an aggregate in bituminous mixtures where its rough texture and relatively high porosity, together with its alkaline activity, produces good adhesion characteristics particularly in the presence of water.

Blast-furnace cements

Table 22 shows that in basic chemical composition slag contains the same elements as Portland cement. It is not a pozzolana nor in itself cementitious, but it possesses latent hydraulic properties which can be developed by the addition of an activator such as lime or another alkaline material.

Many countries use ground granulated blast-furnace slag (GGBFS) in the manufacture of so-called blast-furnace cement (for example, BS 146 (1973),

Table 24. Energy content of hydraulic binders (OECD 1984)

	Energy content as produced (MJ/t)
Ordinary Portland cement (97% clinker)	5000
Granulated slag	40
PFA	25
Cement with high slag content (15% clinker)	1900
Cement with low slag content (65% clinker)	3800

BS 6699 (1986), ASTM C-595-76 (1976)). The percentage of slag in the Portland cement clinker, with which it is ground, may be very low or as high as 85%. The addition of materials such as blast-furnace slag or PFA to Portland cement, even in high proportions, does not impair its hydraulic properties. There may be a decrease in the early strength of concretes made from such cements but the long-term strength is not affected and the incorporation of slag or PFA can give rise to considerable savings in the energy consumed in the production of hydraulic binders (Table 24).

The energy requirements for the production of Portland cement are very high in relation to those for other components of concrete and cement-bound materials. It follows, therefore, that the use of slag and pozzolana cements can lead to significant savings in energy. A survey of the consumption of energy in road building by the OECD (1984) showed that a cement-bound gravel stabilized with $3 \cdot 5\%$ of Portland cement and with $3 \cdot 5\%$ of a high-slag cement consumed 268 MJ/t in the first case and 157 MJ/t in the second.

As an alternative to the addition of blast-furnace slag at the works, GGBFS is also available for addition to concrete made on the site. GGBFS is cheaper than Portland cement, so economies can be made.

Lime—blast-furnace slag mixtures

Blast-furnace slag may be added to Portland cement clinker to manufacture Portland blast-furnace cement. This cement may be used for stabilization in the same proportions as ordinary Portland cement, and the properties of the stabilized material are much the same whether ordinary Portland cement or blast-furnace cement is used. Apart from its use in this manner, the cementitious properties of blast-furnace slags have been developed in France under the title of 'grave-laitier' (gravel-slag) to stabilize gravel and sands for sub-base and base construction.

Granulated slag does not possess any hydraulic properties until it is activated. In the grave-laitier process this activation is achieved by small additions of hydrated lime. The reactivity of the slag varies between works and is measured by a so-called alpha (α) coefficient which is defined as

$$\alpha = S \times P \times 10^{-3}$$

where S is the Blaine specific surface (cm^2/g) of the natural fines of the granulated slag ($< 80 \ \mu m$); and P is the friability properties determined by mixing in a ball-mill 500 g of granulated slag with 1950 g of porcelain beads (diameter 18−20 mm) and subjecting the mix to 2000 revolutions at a rate of 50 rev/min. The tests are carried out and the factor P, defined as a percentage finer than 80 μm, is determined after sieving and washing.

On the basis of this coefficient, granulated slags are divided into four classes:

- Class 1: $\alpha < 20$,   not used in road construction
- Class 2: α 20−40, the most frequently used
- Class 3: α 40−60, reserved for materials that are difficult to handle
- Class 4: $\alpha > 60$,   used only exceptionally.

In France, grave-laitier is produced in mixing plants: it consists of a mixture of gravel with 15−20% of granulated slag together with 1% of hydrated lime (as the activator) and a moisture content of 10%. The strengths obtained from grave-laitier are only half those that would be obtained if Portland cement were used. However, granulated slag is cheaper than cement and grave-laitier has considerable advantages, which are summarized (OECD 1977) below, over cement-stabilized gravel as a construction material.

(a)   A relatively large quantity of granulated slag facilitates a homogeneous distribution of the binder in the mass. Part of the slag remains available, enabling a renewed setting (self-healing) should cracking occur.

(b)   Grave-laitier takes a relatively long time to set, allowing several days of storage without difficulty. It also allows flexible organization of the road works, each machine operating individually and at its maximum output.

(c)   Roadworks equipment can be allowed to circulate over the grave-laitier as soon as it is laid. Post-compaction due to traffic is good. The material is suitable for strengthening purposes while traffic is maintained.

(d)   In the case of heavy rain, excess water is simply allowed to drain off before proceeding with compaction. If necessary, materials may be respread, allowed to dry out and then recompacted.

(e)   The setting process is halted under frost, but setting recurs once normal temperatures are reached.

(f)   Strength takes a long time to build up fully (one year or longer) and is not affected by an initial delay in setting.

(g)   The slow rate of setting allows the moduli of the grave-laitier layers to increase progressively with the consolidation of the subgrade and increasing traffic.

Grave-laitier is the most widely used road base material in France and it is estimated that 65% of French roads have a pavement layer composed of grave laitier. An extension of the process is to use air-cooled crushed slag as the coarse component of the mixture when, to use French terminology, the 'grave' is replaced by slag and the product is known as 'grave-laitier tout laitier'. A useful summary of French procedures on the use of hydraulic binders in road construction is available in both French and English (Ministère des Transports 1980).

A similar technique to the 'grave-laitier' process is used in South Africa where GGBFS is known, somewhat confusingly, as milled granulated blast-furnace slag (MGBS). South African specifications (NITRR 1986) give a ratio of four parts of MGBS to one of hydrated lime as the optimum proportions, but suggest that equal parts of MGBS and lime are often used since this is a convenient ratio in practice even though more lime may be used than is needed.

Instead of using air-cooled blast-furnace slag as the coarse component, phosphoric slag may be used. Phosphoric slag is the by-product of phosphorus manufacture but it has a similar chemical composition to blast-furnace slag. A blend of 85% phosphoric slag and 15% granulated blast-furnace slag is used as a road base material in Holland and is imported from there into South-East England where it has been similarly successful (Kent County Council 1985). When slag is used as the coarse component of the blend the hydraulic properties of the granulated slag do not require activation by the addition of basic activators such as lime.

7. Slate waste

Occurrence

North Wales has always been by far the largest producing area of slate although it is also quarried in Mid-Wales, Cornwall, Devon and Cumbria. Slate quarrying reached its peak in the 19th century when the development of the canal and then rail networks meant that slates could be transported cheaply from the slate-producing areas to all parts of the country for use as a roofing material. From the turn of the century slate quarrying fell into rapid decline as tiles once again became the cheapest method of roofing. During the last ten years there has been an increase in production but this is still at a very low level compared with Victorian times.

Only the rock suitable for splitting is acceptable because the main end use is for roofing. This factor together with the losses incurred during cutting and splitting of large blocks of good slate leads to a high proportion of waste at all stages; overall the proportion of waste-to-slate averages about 20:2. A comprehensive account of the production and uses of slate waste has been published by Watson (1980).

The total annual production of slate waste is estimated to be 5·7 million tonnes (Whitbread *et al.* 1991). This is added to the estimated stockpile of about 440 million tonnes, which puts it second only to colliery spoil in terms of the amount of waste available. However, unlike colliery spoil the stockpiles are highly concentrated in small, remote areas of the country (Table 25).

Table 25. Location of slate waste stockpiles (Whitbread et al. *1991)*

Location	Stockpile (million tonnes)
North Wales	350
Elsewhere in Wales	15
Lake District	15−20
Cornwall/Devon	5−10
Scotland	50
Total	435−445

Composition

The nature of the slate waste varies according to its origin. Mill waste, mainly the ends of slate blocks and the chippings from the dressing of the slate, consists mainly of slate itself, but rocks such as cherts, which are sometimes interbedded with the slate and igneous rocks may also be found in it.

Slate, which forms the major portion of the waste, is a fine-grained aggregate of chlorite, sericite, quartz, haematite and rutile (see Table 26).

A typical chemical analysis of slate waste is given in Table 27. The waste is chemically inert and is most unlikely to cause any chemical problems when used in road construction.

A summary of the physical properties of Welsh slate waste is given in Table 28. This shows that slate waste, provided that it can be crushed to satisfy the relevant grading requirements will satisfy most other requirements with regard to factors such as plasticity, particle strength and durability.

Uses in road construction

Slate waste is, in effect, a crushed rock and is therefore potentially suitable for all applications where crushed rock is specified. Evidence for this is provided by the fact that it is used as a Type 1 granular sub-base material in those areas of North Wales where it occurs (Mears 1975) (see Fig. 27). It is thus apparent that it could also be used for bulk fill, as a selected granular fill material and as a granular capping material. With further processing and screening it would have wider use as an aggregate provided the usage did not have restrictions on the flakiness which is characteristic of the material.

Within its production area slate waste is used to manufacture a wide range of aggregates for use in road building and ancillary works and the product dominates the local market for road aggregates. However, the annual amounts used average out at only 250 000 tonnes, which is less than 5% of the current production of waste. Slate waste as a sub-base material has been shipped by sea to South-East England on a trial basis but although the material proved to be satisfactory the freight costs were marginally too high for it to be competitive.

Table 26. Composition of slate waste (Crockett 1975)

Component	Composition (%)
Sericite	38–40
Quartz	31–45
Chlorite	6–18
Haematite	3–6
Rutile	1–2·5

Table 27. Typical chemical analysis of slate waste (Crockett 1975)

Component	Composition (%)
SiO_2	45–65
Al_2O_3	11–25
Fe_2O_3	0·5–5·7
K_2O	1–6
Na_2O	1–4
MgO	2–7
TiO_2	1–2

Table 28. Physical properties of slate waste aggregates (Goulden 1992)

Property	Source					
	Penrhyn	Ffestiniog	Llechwedd	Croes-y-Ddu-Afon	Aberllefni	Burlington
Water absorption (%)	0·2	0·3	0·3	0·3	0·2	0·3
Flakiness index (mean)	93	100	100	100	93	98
Elongation index (mean)	23	29	34	34	23	27
Aggregates crushing value (kN)	25	29	26	30	24	23
10% fines value (dry) (kN)	160	130	140	120	170	160
10% fines value (soaked) (kN)	110	90	80	70	110	100
Aggregate impact value	27	29	29	33	28	28
Relative densities						
Oven-dry	2·80	2·76	2·77	2·75	2·80	2·80
Saturated surface dry	2·82	2·78	2·78	2·77	2·82	2·81
Apparent	2·84	2·79	2·80	2·79	2·84	2·83
$MgSO_4$ soundness	99	98	98	98	99	98
Plasticity index	0	0	0	0	0	0
Slake durability index (%)	96	94	95	94	96	96
Sulphate content (g.SO_3/litre)	0·01	0·01	0·01	0·01	0·01	0·01

Fig. 27. Processing of slate waste to produce granular sub-base

8. Construction and demolition wastes

Occurrence

Construction and demolition wastes are materials such as concrete, masonry, bituminous road materials, etc., arising from the demolition of buildings, airfield runways and roads. In 1990 it was estimated that 24 million tonnes of these demolition and construction wastes and 7 million tonnes of asphalt road planings were produced. All the asphalt planings and 11 million tonnes of the other wastes were re-utilized (Whitbread *et al.* 1991). Apart from the 31 million tonnes of construction wastes, this survey also included 46 million tonnes of other construction wastes mainly arising from excavations during building work.

A more recent survey (Howard Humphreys 1994) indicated that about 63% of the demolition and construction wastes which arise is recycled in one form or other. However, the quantities which were recycled to secondary aggregate constituted only 4% of the estimated total arisings. This was because the majority of material which is recycled is employed in either low-level uses close to the site of arising (about 30%), or to landfill engineering (also about 30%). The remainder was not recycled at all and was presumably used as inert landfill material.

Waste from road reconstruction forms only one quarter of the total amount of construction waste produced. However, it clearly has a high potential for re-use in roadmaking as, not only does it occur on site, but it is likely to retain at least some of the properties that originally made it suitable for use in road construction. The recycling of construction materials has long been recognized to have the potential to conserve natural resources and to reduce the energy used in production. In some countries it is a standard alternative for both construction and maintenance, particularly where there is a shortage of road-building aggregate. In the UK a plentiful supply of good quality aggregate, relatively short distances between quarry, mixing-plant and site and a wide range of specifications have hitherto reduced the need for recycling.

However, the climate of opinion is rapidly changing because of increasing concern to conserve natural resources and reduce the environmental impact of road construction. As a result, a considerable amount of research has been carried out recently into methods of re-using construction materials, particularly those arising from the reconstruction of roads. This work has led to the inclusion in the Specification for Highway Works (1991a) of the option of using recycled materials as a replacement, either in whole or in part, for natural aggregates.

Composition

Apart from asphalt road planings, Mulheron (1991) distinguished four main categories of construction and demolition wastes. These are, in order of potential use:

(*a*)   clean crushed concrete: crushed and graded concrete containing less than 5% of brick or other stony material

(*b*)   clean crushed brick: crushed and graded brick containing less than 5% of other material such as concrete or natural stone

(*c*)   clean demolition debris: crushed and graded concrete and brick

(*d*)   crushed demolition debris: mixed crushed concrete and brick that has been screened and sorted to remove excessive contamination, but still containing a proportion of wood, glass or other impurities.

Crushed concrete

Any crushed aggregate produced from construction and demolition wastes tends to be referred to as crushed concrete (O'Mahoney 1990), but the term should be reserved for crushed concrete produced from the break-up and crushing of concrete slabs from road and airfield pavements. Concrete is also available from the demolition of buildings but it is likely to be reinforced, which makes it difficult to crush, and contaminated with other building materials.

Crushed concrete arising from the demolition of disused airfield runways was widely available after World War 2. This source gradually declined but may temporarily regain importance as, following the end of the Cold War, further airfields will probably be closed. Concrete from this source and from pavement-quality concrete removed from roads can, provided it does not contain reinforcement, be crushed to produce an aggregate which can be assumed not to contain any harmful components. Any material containing other constituents as well as crushed concrete, e.g. brick, glass, asphalt etc., would, under Mulheron's classification (*a*)−(*d*) above, go into the category of crushed demolition debris.

Unfortunately there is no generally recognized method of classifying crushed concrete so that it can be readily distinguished from other construction and demolition wastes. A Dutch classification for recycled granular base materials described by Sweere (1991) defined crushed concrete in the following manner.

(*a*)   Main components: at least 80% by weight of crushed gravel or crushed aggregate concrete; at most 10% by weight of other broken stony material, the particles of which shall have a particle density of at least $2 \cdot 1$ Mg/m^3.

(*b*)   Additional elements: at most 10% by weight of other crushed stone or stony material. As for broken asphalt, this shall not exceed 5%.

(*c*)   Impurities: at most 1% non-stony material (plastic, plaster, rubber etc.); at most $0 \cdot 1$% decomposable organic matter such as wood and vegetable remains.

Crushed brick

The fate of demolished brickwork depends both on the type of brick and the type of mortar. It is often economic to recover the bricks, for which there is a significant second-hand demand, but cheaper bricks such as flettons would usually be sold as hardcore. The type of mortar used will influence the decision; lime mortar is easily separated from the bricks but cement-containing mortars are very difficult to remove. Contamination from either lime or cement mortar is not likely to be a problem but contamination from gypsum plaster could result in the crushed bricks having unacceptably high sulphate contents. Even without contamination from gypsum some bricks have soluble sulphate contents high enough to be deleterious.

Crushed demolition debris

Crushed demolition debris other than crushed concrete and crushed brick has no applications for use except as general bulk fill. Provided that it meets the requirements given earlier there is no reason why it should not be used for bulk fill, but care needs to be taken with its use because it can be very heterogeneous. Rubble containing timber should be avoided because, when it rots, cavities will be left in the fill.

Bituminous materials

Bituminous materials available from road reconstruction and maintenance fall into two categories: one, asphalt road planings, can be truly classified as a demolition waste; the other, concerned with the in-situ recycling of the road surface, cannot because the material never leaves that road and there is no stage at which the material is available for other potential use.

Recycling of construction and demolition wastes

Bituminous materials

Significant savings in energy consumption can be made by recycling bituminous materials removed during maintenance. There are several methods of recycling bituminous pavement layers. They have been reviewed by Mercer and Potter (1990) and can be categorized as: in-situ recycling, where processing takes place without transporting the reclaimed material; and central plant recycling, in which the excavated material is taken to another location for treatment. Both processes can be further sub-divided into hot and cold recycling.

Due to the potential savings in energy consumption that can be gained by the use of recycling techniques a good deal of attention has been paid to this subject recently. Table 29 shows the estimated energy consumption on a road project in Kent in the production of raw materials for central plant recycling of hot-rolled asphalt using different percentages of reclaimed bituminous planings.

Methods and equipment for recycling old bituminous pavements into new pavements are well developed and widely used. According to a review by the OECD (1977) the benefits of such recycling are:

- reduced transport costs and fuel requirements
- reduction in aggregate requirements and elimination of potential disposal problems

Table 29. Energy consumption in the production of raw materials for asphalt (Hubert 1987)

Material	Energy (MJ/tonne asphalt produced)		
	0% recycle	40% recycle	60% recycle
Sand	9·1	3·6	1·2
Crushed aggregate	27·7	20·0	15·6
Bitumen	75·1	47·3	27·8
Planings	0·0	8·5	12·7
Total	111·9	79·4	57·3
Energy savings (%)	0·0	29·0	48·8

- up to 75% reduction in bitumen content required
- reduction in fuel required for drying
- generally lower emissions during construction.

Recycling of bituminous materials may involve the re-use of asphalt road planings or the in-situ renovation of the surfacing.

Asphalt Road Planings. Asphalt road planings are the bituminized roadstone which is removed from the surface of roads prior to maintenance resurfacings or full reconstruction. For resurfacing only a 40 mm layer of the road surface is planed off by machinery designed specifically for the purpose. When a road is to be completely rebuilt, the entire bituminous layer, which may include the road base, is removed. The reconstruction may also involve the removal of cement-bound and unbound base and sub-base materials, the use of which is considered below.

Whitbread *et al.* (1991) showed that road planings were rarely wasted and many alternative uses were found, although this did not include the off-site recycling of planings to produce fresh bituminous material. However, research by Cornelius and Edwards (1991) has shown that planings can be recycled to produce material which can comply with the relevant specifications when it is produced using modified drum-mixers and laid using conventional plant.

As a result of this and other work, the latest edition of the Specification for Highway Works (1991a) recognizes that road planings have a high potential for re-use, and it contains a new clause (clause 902) which specifies that 'up to 10% of materials derived from planing of existing carriageways may, with the approval of the Engineer, be used in the production of bituminous materials'. This differs from the 1986 edition which stipulated that 'Surplus materials arising from the planing off process shall be disposed of by the Contractor'.

In-situ hot recycling is used for rehabilitating wearing courses. Two complementary processes are available and are known, respectively, as Repave and Remix.

Repave restores surfaces that are in sound structural condition by bonding a thin overlay or inlay to the preheated, scarified and reprofiled road surface. It can be considered to be recycling because the existing wearing course is re-used in combination with a thin layer of new material. Since 1982 the process has been approved by the Department of Transport and it was included in the 1986 edition of the Specification for Highway Works as a maintenance treatment for trunk roads and motorways. Repave is not recommended for surfaces that show signs of structural weakness or cracking, which generally indicates defects in the material that could be exacerbated by heating the surface.

The Remix process can be completed in a single pass of a purpose-built machine which preheats and scarifies the surface before augering it into a pugmill where it is blended with freshly-mixed new material. This recycled mix is placed evenly on the heated surface to form the replacement wearing course. Existing wearing courses that have deformed or become brittle may, within limits, be modified to comply with existing specifications and thus overcome these deficiencies. The composition of the old surface has to be determined so that the added virgin material can be selected to produce a blend that meets the specification requirements.

Cold in-situ recycling restores the structural integrity of the road by pulverizing it to a depth of up to 350 mm, mixing in cement, bitumen emulsion or foamed bitumen, and compacting. A new wearing course is then applied which can vary from a double surface dressing to a 100 mm overlay, depending on anticipated traffic level. Site investigation is necessary to decide if cold in-situ recycling is technically feasible and, if so, to select the depth of pulverizing and the type of binder. Cold in-situ processes are more applicable to lighter-trafficked roads.

Crushed concrete

The Specification for Highway Works (1991a) allows the use of crushed concrete as a substitute for natural aggregates for most purposes. However, although not specifically stated, it can be assumed that, where it is allowed, the specification envisages it will relate to crushed concrete as defined above and not to demolition debris. This is made particularly apparent in those instances where the quality of the crushed concrete is critical to the performance of the material to be produced from it. Thus, when crushed concrete is to be used as the aggregate in cement-bound road base materials CBM3 and CBM4 and for pavement or for pavement-quality concrete, the specification requires that the crushed concrete shall comply with the grading requirements of BS 882 (1992).

Crushed cement-bound sub-base and road base

Some design procedures envisage a cement-bound road base which at the end of its useful life is removed and replaced (Department of Transport 1987). For whatever reason, the removal of long sections of cement-bound road base or sub-base can give rise to large quantities of material which may have potential for re-use. For example, they may be crushed to provide a good quality granular material which meets all the requirements for granular sub-base materials (Fig. 28).

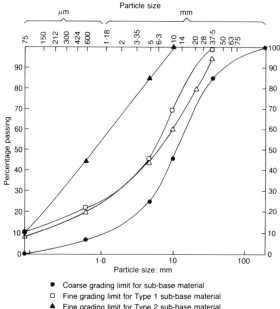

● Coarse grading limit for sub-base material
□ Fine grading limit for Type 1 sub-base material
▲ Fine grading limit for Type 2 sub-base material
△ Grading of crushed CBM3 road base

Fig. 28. Grading curves of crushed CBM3 road base (Sherwood 1993)

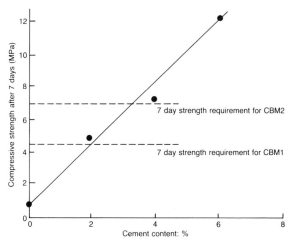

Fig. 29. Relation between cement content and strength for crushed CBM3 road base material (Sherwood 1993)

The granular material, produced by crushing, may also be re-stabilized with cement to produce a cement-stabilized material in a lower category. Fig. 29 shows the relation between cement content and strength for a crushed CBM3 road base. This shows that it can easily be made to meet the requirements for CBM2 sub-base and that the material may have residual cementitious properties as a result of the exposure of unhydrated cement during crushing.

9. Other materials

Spent oil shale

Occurrence

Spent oil shale is the waste material from the now defunct oil extraction industry which was concentrated in the West Lothian area of central Scotland. At its peak in 1913 the industry employed 13 000 men and produced 3·3 million tonnes per year of waste (Burns 1978). Production steadily fell and it became increasingly difficult for the industry to compete with imported oil from the Middle East.

Production finally halted in 1962 by which time a large stockpile of spent shale had accumulated. This was estimated (Department of the Environment 1991a) at 100 million tonnes occupying 395 hectares of land in a small area around Livingston and Bathgate, about 20 km to the west of Edinburgh.

Composition

Following the extraction of the crude oil and naphtha from the oil shale, the spent shale, together with other materials considered unsuitable for processing, was deposited in heaps on land adjacent to the refineries and mines. Spontaneous combustion within the tips sometimes caused further changes to occur.

Although of different origin, spent oil shale is not unlike burnt colliery spoil in its chemical and physical properties. Like burnt colliery spoil it is pinkish in appearance and samples may be obtained with a particle size distribution corresponding to that required for granular sub-bases (Fig. 30). A typical analysis, compared with an analysis of burnt colliery spoil, is given in Table 30. The results of sulphate and loss on ignition determinations of a range of spent oil shales are given in Table 31.

The range of sulphate contents found in spent oil shale is much the same as the range found in burnt colliery spoil (Tables 9 and 10). For this reason the chemical problems associated with its use are also very similar and the earlier discussion on this issue of colliery spoil applies equally to spent oil shale. Spontaneous combustion is clearly not a problem because the shale was heated to extract the oil. Neither is the presence of sulphides because these will have been driven off or converted to sulphates during the extraction process.

Uses of spent oil shale in road construction

Due to its close similarity to burnt colliery spoil, spent oil shale may be used for all purposes where burnt colliery spoil is permitted to be used. It has

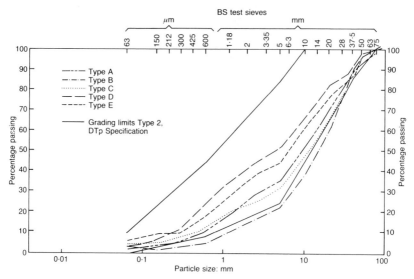

Fig. 30. Particle size distribution of spent oil shale samples (Burns 1978)

Table 30. Comparison of the chemical composition of burnt colliery spoil and spent oil shale

Component	Burnt colliery spoil (%)*	Spent oil shale (%)†
SiO_2	45−60	48·5
Al_2O_3	21−31	25·2
Fe_2O_3	4−13	12·1
CaO	0·5−6	5·3
MgO	1−3	2·2
Na_2O	0·2−0·6	NR
K_2O	2−3·5	NR
SO_3	0·1−5	3·2
Loss on ignition	2−6	3·0

* Range of values (Sherwood 1987)
† Typical analysis (Burns 1978)

been widely used as a bulk fill material in central Scotland with very good results (Fraser and Lake 1967), and this is the main outlet for its use in road construction. Fig. 30 shows that its particle size distribution may meet the grading requirements for granular sub-base materials, and the Specification for Highway Works (1991a) permits its use. However, as in the case of burnt colliery spoil, the recent imposition of particle strength and soundness requirements makes it unlikely that it would conform to all the requirements of the specification.

Table 31. Results of sulphate and loss on ignition determinations on selected spent oil shales (Burns 1978)

Sample no.	Total sulphate (% as SO_3)	Loss on ignition (%)
1	2·20	1·87
2	2·80	1·25
3	0·70	1·35
4	2·40	2·26
5	2·80	1·07

Steel slag

Occurrence

The manufacture of steel involves the removal from the iron of excess quantities of carbon and silicon by oxidation and the addition of small quantities of other constituents that are necessary for imparting special properties to the steel. There are three main types of steel-making furnace that can be used: open hearth, basic oxygen (BOS) and electric arc (EAF), but only BOS and EAF slags are now produced in the UK. Table 32 gives the tonnage and utilization of the two types of slag produced.

Composition

When air-cooled steel slags give a product that resembles igneous rock. They have a particle density about 1·25 times higher than blast-furnace slag and may also be mechanically stronger and more durable. Unfortunately, however,

Table 32. Production and utilization (in milliom tonnes) of steel slag in 1989—90 (Whitbread et al. 1991)

Location	Amount produced	Utilization in 1989—90	Current stockpile
BOS slag			
Ravenscraig	0·30	0	2
Teesside	0·35	0	2
Scunthorpe	0·40	0·16	5
Llanwern	0·30	0	2
Port Talbot	0·30	0	1
Total	1·65	0·16	12
EAF slag			
Rotherham	0·25	0·25	0
South Wales	0·12	0	0·1
Sheerness	0·08	0	0·1
Total	0·45	0·25	0·2
Grand total	2·10	0·41	12·2

they may contain residual iron, free lime (CaO) or free magnesia (MgO). If present, the hydration of lime and magnesia makes steel slags unstable and liable to expand.

The hydration of calcium oxide is rapid but it may be locked up within the slag particles and the rate of reaction is much reduced, while the hydration of magnesium oxide is slow even under favourable conditions. Both hydration processes are accompanied by an increase in volume which can cause disintegration of the slag particles over a considerable period of time.

At present there are no UK specifications relating to the use of steel slag. Verhasselt and Choquet (1989) gave the following recommendations for Belgian steel slags:

(a) steel slag should not have a free lime (CaO) content of more than 4·5% at the time of production
(b) before use the slag should be allowed to weather for one year
(c) the maximum particle size should not exceed 20−25 mm
(d) before use the volumetric stability should be checked by a volumetric swelling test.

The possibility of setting limits on the free lime content of steel slag is also being considered for inclusion in the draft European Standard for Hydraulic-bound and Unbound Aggregates. It is intended that this should set limits on the free CaO content of steel slags and also include an accelerated expansion test (CEN 1992).

Uses of steel slag in road construction

Although in many respects steel slags appear to be good-quality roadmaking aggregates, the possible presence of free lime (CaO) and free magnesia (MgO), with the consequent risk of expansion if and when these hydrate, severely limits their use in road construction and virtually excludes them from use as fill under structures.

The questionable volumetric stability of steel slag explains why so little is used (Table 32); less than 10% of the current production of BOS slag is used and there is a stockpile of 12 million tonnes. EAF slag has a lower free oxide content and about half of the current production is used. It is processed by crushing, grading and screening to make road aggregates. The processing is confined to the slag arising in the aggregates. The processing is confined to the slag arising in the Rotherham/Sheffield area because the volumes available elsewhere are insufficient to justify a processing plant.

Due to the possibility of long-term expansion steel slags are used, if at all, only in those situations where expansion is unlikely, as in the case of dense bitumen macadam, or where any expansion that does occur is not likely to be a serious problem, as in the case of surface dressing. Their main use, therefore, is in the upper bituminous layers of the road structure where the fact that the material is impermeable and the aggregate particles are coated with bitumen means that water would have difficulty in penetrating into the particles to cause any hydration. If any expansion did occur it would be limited to the upper layers which would cause less serious disruption than expansion occurring in the lower layer.

Incinerated refuse

Occurrence

According to the Royal Commission on Environmental Pollution (1993) incineration is an efficient way of recovering the energy present in household wastes in order, for example, to generate electricity. It can help significantly in countering the greenhouse effect because it prevents the wastes from decaying and producing methane gas. Compared with untreated household wastes, the residues left after incineration take up less space and are much more stable. Incineration of household wastes is therefore environmentally preferable to tipping them, into a landfill site.

Direct incineration results in a 60% reduction in weight with a 90% reduction in volume. Half of the incinerated refuse is ferrous metal which can be magnetically extracted and recyled, while the other half is ash. In 1976, 33 incineration plants were operational in England (Roe 1976) which produced between them 1 million tonnes per year of ash. At that time more were planned, but problems with some plants led to their closure. An estimate of waste disposal statistics in 1988 (CIPFA 1988) gave a total of 26·8 million tonnes of refuse being disposed of by waste disposal authorities of which only 1·54 million tonnes (6%) was being incinerated.

The CIPFA estimate that 6% of household waste was being incinerated is in broad agreement with a more recent estimate of about 7% by the Royal Commission on Environmental Pollution (1993). Using the CIPFA figures and assuming a 60% reduction of mass on incineration, this would give rise to 616 000 tonnes of incinerated residue, half of which (313 000 tonnes) would be ash. As the amount of waste being produced and the proportion that is incinerated have not fallen between 1988 and 1993 this calls into question the estimate of only 160 000 tonnes per year in 1990, with one plant (Edmonton) accounting for 100 000 tonnes of the total production (Whitbread *et al.* 1991).

The Royal Commission on Environmental Pollution concluded that in the short term the proportion of waste that is incinerated may well decrease because existing plants cannot meet new standards for emissions. However, it concluded that, on environmental grounds, incineration of domestic refuse was preferable to its use for landfill and 'it offered the best practical environmental option for disposing of the increasing level of household waste'. It is possible, therefore, that the production of incinerated refuse will increase in the future. Impetus for this may come from the announcement by the Government in late 1994 that it intends to impose a tax on waste dumped in landfill sites.

Composition

Incinerated residues consist mainly of clinker, glass, ceramics, metal and unburnt matter. The metal is largely in the form of tin cans and most plants magnetically extract the ferrous metal before discharging the residues. The unburnt matter may be paper, rag or putrescibles, the amounts depending on various factors including the type of furnace, the temperature of the firebed or the length of time the materials spend in it, and on the composition of the raw refuse itself. A complete chemical analysis of the Edmonton ash is given in Table 33 and the properties of some other ashes are given in Table 34.

Table 33. Chemical analysis of the
inorganic constituents of ash from
Edmonton (Roe 1976)

Component	Percentage present
SiO_2	35
Fe_2O_3	26
Al_2O_3	23
CaO	6·3
Na_2O	2·5
ZnO	1·2
SO_3	1·0
MgO	0·7
TiO_2	0·6
MnO	0·3
Other	3·2

Table 34. Chemical properties of some UK ashes (Roe 1976)

	Edmonton	Nottingham	Derby	Birmingham	Sunderland
Loss on ignition (%)	2·6	5·7	4·2	6·4	15·2
Soluble sulphate (as g.SO_3/litre)	2·32	3·15	2·70	5·51	4·32
Total sulphate (as % SO_3)	0·60	1·36	0·67	0·95	1·09
pH value	9·4	10·0	8·7	7·6	9·9

Uses of incinerated refuse in road construction

Although some ashes may have gradings that make them potentially suitable as a selected granular fill or unbound sub-base material (see Fig. 31), the main use of incinerated refuse is for bulk fill. It cannot be used as a cement-bound material because it may contain aluminium, heavy metals and glass, all of which are capable of reacting adversely with cement.

Non-ferrous slags

Apart from blast-furnace slag and steel slag, which have already been considered, small quantities of slag are produced from the smelting of other metal ores which may find an outlet in roadmaking in the vicinity of the works. The amounts available of such slags have been estimated (Whitbread *et al.* 1991) as:

- Copper slag: 20 000 tonnes/year
- lead−zinc slag: 100 000 tonnes/year
- tin slag: 60 000 tonnes/year.

Little is known of the potential uses of these slags in road construction and

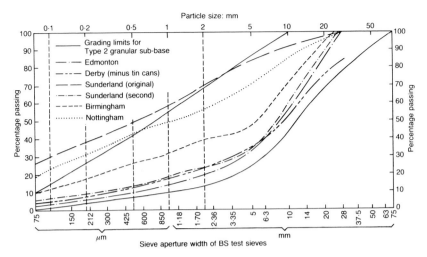

Fig. 31. Comparison of particle size distributions of incinerated refuse from various sources (Roe 1976)

the amounts produced in this country are too small to justify much research on their properties. They are produced in very much larger quantities in the USA and have been used in road construction to a limited degree (OECD 1977).

The main problem with such materials is that they may contain components that are unstable in the presence of water and/or that water leaching through them may contain harmful pollutants.

Imported materials

Most waste materials and industrial by-products are of low value and are used only in the areas where they occur. It is therefore extremely unlikely that any would be imported into this country from the continent. The only exception to this is the importation of slags into South-East England which occurs for three reasons:

(*a*)   the shortage of good quality aggregates in the South-East

(*b*)   slags, in general, are high quality aggregates which can command a price high enough to make transporting them an economic proposition

(*c*)   compared with the UK, slag production in neighbouring countries is much higher.

Slags produced on the continent do not differ substantially from those produced by similar processes in this country.

10. Conclusions

Suitability of alternative materials

The review of the use of wastes and by-products in road construction in Part 2 has shown that all the mineral wastes and by-products currently available in the UK have some potential uses and that specifications do not unreasonably restrict their use. Table 35 summarizes the conclusions that can be drawn with regard to the materials considered.

Selection of material to use

The environmental aspects of deciding whether or not to use an alternative to naturally-occurring materials are considered in Part 3. However, assuming that an alternative material is readily available at economic cost, a decision still has to be made on whether or not it is suitable. The information given in Parts 1 and 2 should enable a decision to be made and this information is summarized in the flow diagrams given in Figs 32—36.

Table 35. Summary of potential uses of waste materials and by-products in road construction

Material	Bulk fill	Unbound capping layer	Unbound sub-base	Cement-bound material	Concrete aggregate or additive	Bitumen-bound material	Surface dressing aggregate
Crushed concrete	High*	High	High	High	High	Some	None
Asphalt planings	High*	High	High	Low	None	High	None
Demolition wastes	High	Some	Some	Low	Low	Low	None
Blast-furnace slag	High*	High	High	High	High	High	High
Steel slag	Low	Low	Low	Low	Low	Some	High
Burnt colliery spoil	High	High	Some	High	Low	Low	None
Unburnt colliery spoil	High	Low	None	Some	None	None	None
Spent oil shale	High	High	Some	High	Low	Low	None
PFA	High	Low	Low	High	High	None†	None
FBA	High	Some	Some	High	Some	Low	None
China clay sand	High	High	Some	High	High	Some	Low
Slate waste	High	High	High	Some	Some	Low†	None
Incinerator ash	High	Some	Some	None	None	None	None

* Suitable but inappropriate (wasteful) use
† PFA and slate dust can be used as a filler

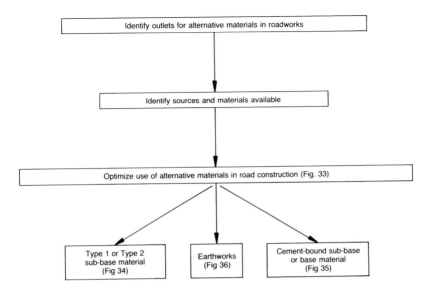

Fig. 32 Flow diagram: deciding whether to use an alternative material (based on BS 6543, 1985a)

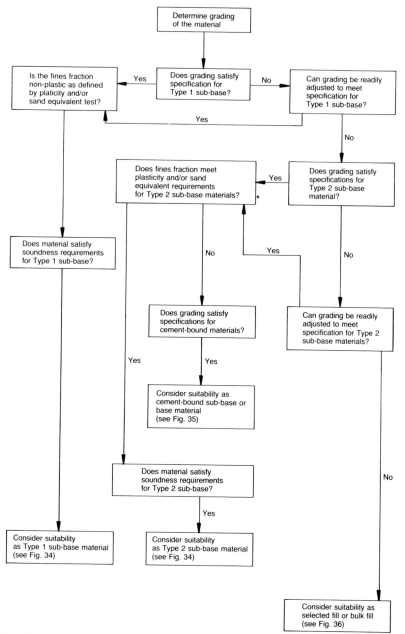

Fig. 33. Optimizing the use of alternative materials in road construction (BS 6543, 1985a)

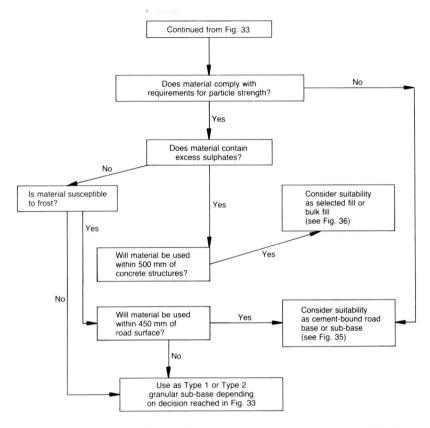

Fig. 34. Determining suitability as Type 1 and Type 2 granular sub-base (BS 6543, 1985a)

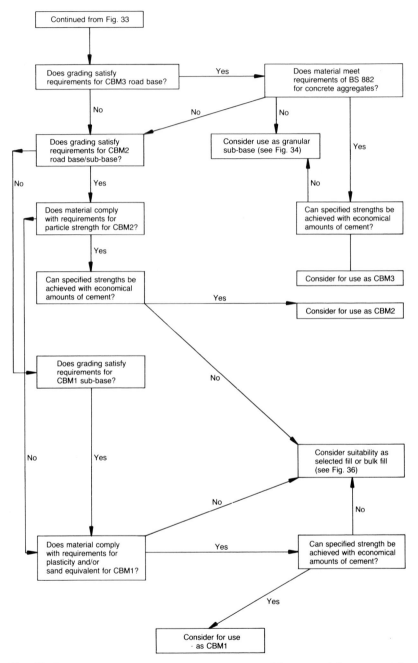

Fig. 35. Determining suitability as cement-bound road base or sub-base material

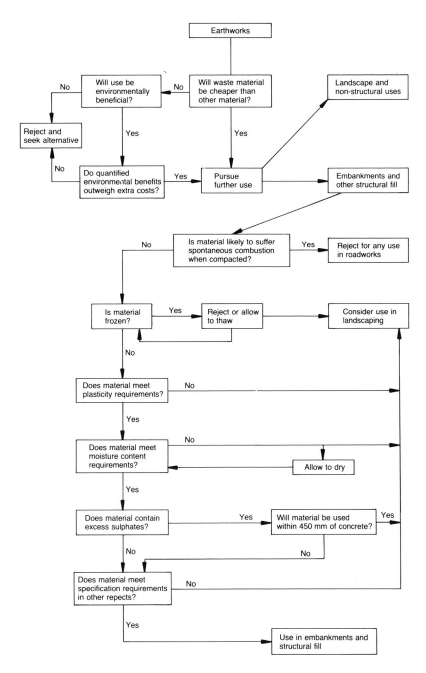

Fig. 36. Determining the suitability for use in earthworks construction (BS 6543, 1985a)

PART 3
ENVIRONMENTAL AND
ECONOMIC CONSIDERATIONS

Before any new road is built a cost-benefit analysis will have been made to ensure that the anticipated benefits to be gained exceed the economic costs and the environmental disturbance resulting from its construction. There is much scope for argument on this subject as can be seen from the prolonged and heated discussions that take place at public inquiries on the routing of new roads. No satisfactory method of assessment has yet been developed that will satisfy all parties on whether or not a road should be built. However, assuming that a decision has been made to build a road, all would agree that it is axiomatic that the materials used for its construction should be obtained at the lowest cost (commensurate with satisfactory performance) and with the minimum of environmental disturbance.

The economic factors relating to the materials used in construction mainly derive from the costs of extracting the material, processing and hauling it to the site, all of which are closely related to energy consumption. The environmental disturbance is made up of factors such as disturbance to the landscape leading to possible dereliction, disturbance caused by transporting material from source to site, and the depletion of natural resources. The three factors of resource depletion, environmental degradation and energy consumption are not completely independent of each other and steps taken to reduce the effect of one may be accompanied by an increase in another, but clearly any project should be planned in such a way that all effects are minimized.

The use of alternative materials, such as those considered in this book, in place of naturally-occurring materials, can help to achieve this aim. Care needs to be taken as the use of an unsuitable material for a particular set of circumstances can lead to a costly failure and create a climate of opinion that is hostile to its use. The primary requirement in selecting a waste material or by-product must, therefore, be that its use in preference to other materials will not shorten the life of the road structure. However, over-specification is wasteful of resources and a balance has to be struck between these two conflicting aims.

Parts 1 and 2 dealt with the materials that are available and the uses which can be made of them in road construction. However, if a material is to be used the economic and environmental factors arising from its use also need to be considered. Part 3 is therefore concerned with a discussion of the factors that should be taken into account in deciding whether or not to use alternative materials, such as waste materials or by-products, in preference to naturally-occurring materials. It also describes the legislation relating to the use of waste materials and by-products in the UK.

11. Environmental effects of aggregate and waste production

The environmental impacts of aggregates extraction are a source of increasing concern in many parts of the country. These impacts have been itemized (CPRE 1993) as loss of mature countryside, visual intrusion, heavy lorry traffic on unsuitable roads, noise dust and blasting vibration. Extraction of aggregates also represents the loss of two finite resources, the aggregates themselves and unspoilt countryside from which they are extracted. The harmful effects of aggregate construction can be considerably ameliorated by the attachment of restoration conditions to the planning consents so that it can be difficult to discern that any disturbance has occurred (compare Figs 37 and 38). However, the Department of the Environment (1991b) admits that 'landscape conditions attached to surface mineral workings have often been ignored ... and the conditions are neither always implemented or enforced'.

In the 30-year period to 1991 the production in the UK of newly-quarried aggregates increased from 110 million tonnes to a peak of about 300 million tonnes in 1988, falling back to about 250 million tonnes by 1991 (see Fig. 1). Despite the drop in demand in recent years demand is expected to increase still further, and the Department of the Environment (1994b) has published forecasts (Fig. 39) which indicated that by 2011 the annual consumption of aggregates in England and Wales (Scotland accounts for an additional 10% of consumption) would increase to about 400 million tonnes. In volumetric terms the amount of aggregates excavated has been likened (Adams 1991) to the equivalent of digging a hole 3·3 times the area of Berkshire ($1·25 \times 10^9$ m^2) to a depth of 6 m in the period 1990−2010.

Road building plays a significant role in the demand for aggregates as it accounts for about one-third of the total production (Fig. 2). In 1989, 96 million tonnes were used and, even if the amounts used by local authorities in local road construction are excluded, it is estimated that the current road building plans of the Department of Transport in the 1990s will use 510 million tonnes. This represents about 255 million cubic metres of aggregate taken from a hole 10 m deep and 2550 hectares in area. The implications of this have been considered by the Royal Commission on Environmental Pollution (1994) which reported that

'We are concerned that extensive damage to the environment would be caused through extraction of the aggregates to carry out the present road building programme. We do not consider that the implied rate of consumption can be regarded as sustainable'.

Fig. 37. Gravel workings, Harlington, Middlesex (1976)

Fig. 38. Restored gravel workings, Harlington, Middlesex (1981)

Raw materials for civil engineering construction have to be obtained by open-cast methods which scar the landscape (Fig. 37) while the mineral wastes for which there is little demand are usually stockpiled in spoil tips (Figs 12 and 20 give examples). In some cases the waste materials can be tipped into the holes produced by mineral excavations; this method is widely used for the disposal of domestic refuse. When this can be done it offers an attractive

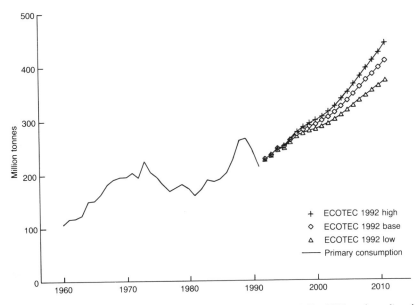

Fig. 39. Consumption of aggregates in England and Wales 1960–1992 and predicted consumption 1992–2011 (Department of the Environment 1994b)

Table 36. Total area (in hectares) of derelict land in England* 1988 (Department of the Environment 1991)

Region	Spoil heaps Colliery	Spoil heaps Other	Excavations and pits	Military dereliction	Derelict railway land	Other	Total
North	967	797	583	282	1256	2060	5945
North-West	703	629	1339	391	1216	4545	8823
Yorkshire/ Humberside	885	390	966	210	1382	2312	6145
West Midlands	1185	335	589	649	590	2227	5575
East Midlands	911	366	655	505	888	1082	4407
East Anglia	0	12	172	179	177	53	593
South-West	36	4483	250	135	616	307	5827
South-East	8	30	1020	76	154	506	1774
Greater London	0	157	390	149	134	556	1386
Total	4695	7199	5694	2576	6413	13918	40495

* Values refer only to England: Scotland and Wales have proportionately more

solution but, unfortunately, it is not always possible to do so. Apart from the fact that tipping into holes can cause pollution of underground water supplies, most waste materials are not produced sufficiently close to holes in the ground for such disposal to be an economic proposition. Consequently, the UK has large areas of derelict land made up of pits and waste heaps (Table 36). The latest estimate of the total amount of derelict land (Department of the Environment 1991) is more than 40 000 hectares, i.e. roughly the area of the Isle of Wight.

The amount of derelict land is, in fact, greater than is indicated by Table 36 because the table only includes land which is defined as being 'so damaged by industrial or other development that it is incapable of beneficial use without treatment'. It does not include active workings, land to which restoration conditions have already been attached, or land which, although originally derelict, is no longer unsightly.

These exclusions go a long way to explaining the discrepancies between the official figures for the extent of the problem and the generally held view that there is much more derelict land about than the official figures suggest. From a layman's viewpoint the general definition suggested by the Civic Trust is probably better: 'Virtually any land which is ugly or unattractive in appearance: spoil heaps, scrap or rubbish dumps, excavations, dilapidated buildings, subsided or any land which is neglected'. To a casual observer the problem is enlarged by the fact that once a certain proportion of land in a given area becomes derelict the land in between, although not derelict by any of the definitions given above, appears to be so — a series of heaps or holes in an area blights the interstices as well.

Parts 1 and 2 have shown that the alternative materials considered in this report can, in many instances, be used as substitutes to the naturally-occurring materials used in civil engineering. These parts were concerned with technical considerations but it is clear that their use can have considerable environmental benefits. These benefits are threefold as they can lead to the conservation of natural resources, the disposal of the waste materials, which are often the cause of unsightliness and dereliction, and the clearance of valuable land for other uses.

This fact has been recognized by the Government which has stated that: 'To reduce the environmental effects of quarrying new materials the Government is keen to encourage the greatest possible use of waste and recycled materials in accordance with the principles of sustainable development' (DOE/DOT 1992), sustainable development being defined (Department of the Environment 1993b) as meeting the need of the present generation without compromising the ability of future generations to meet their own needs. The implications of sustainable development for minerals planning are that avoidable and irretrievable losses of natural resources should be limited. This means making the best and most efficient use of all available resources. It is thus clear that alternative materials, many of which occur in large quantities, can play an important role in achieving this aim.

Unfortunately the environmental benefits may be offset by increased economic costs and, even if an alternative material is technically suitable, there may be other considerations which mean that it is unlikely to be used. The problem was concisely summarized by the Advisory Committee on Aggregates (1976) in its reference to china clay sand:

'China clay sands need only the same basic grading and washing processes that are applied to natural aggregates. Much is readily available in heaps that are largely free of contamination by overburden and other deleterious matter. It is there virtually for the taking. To use it would conserve resources in other parts of the country, and permit land which might be used for aggregates construction to be used for other purposes. The continued existence of spoil heaps imposes costs on producers and the community alike — sterilizing further mineral development, destroying agricultural land and causing visual and other disbenefits. However, the material is located far from main centres of demand and the cost of transport by rail or sea has not proved viable. Thus we have the apparently nonsensical position that holes are being dug in one part of the country and heaps of sand being raised in another. The problem can only be solved on a national basis and the gap can only be closed by a national subsidy justified for environmental reasons. It is a sad commentary on our value judgement that the problem clearly has a lower priority than Concorde'. (Note: the Concorde project involved the expenditure of £2000 million (sic), at 1976 prices; only 15 aircraft ever flew commercially and even these had to be given away because no airline would buy them).

The dilemma remains; how these opposing factors can be balanced so that an optimum decision can be reached forms the subject of the remainder of this Part 3.

12. Decision making — traditional or alternative materials?

The advantages to be gained by using alternative materials in place of naturally-occurring materials are finely balanced and it is rarely self-evident as to which is the optimum solution. No generally accepted method of making the decision has evolved but this chapter discusses the factors that have to be taken into account; it is based on a methodology originally put forward by Sherwood and Roe (1974) and further developed by the OECD (1977).

Advantages of using alternative materials
Benefits of removal of waste tips
The use of a tip as a source of material for road construction is an attractive proposition, but it is seldom feasible to remove the whole of the tip. The limited removal of material may hinder rather than help eventual restoration; if only partial/removal of a deposit is planned then the stability and landscaping of the remainder will require consideration to avoid aggravating dereliction.

It is important to bear in mind that finding a beneficial use for the material in a tip is not the only satisfactory solution as it may also be re-contoured and vegetation established on its slopes, as in Fig. 40, so that it is no longer an eyesore and blends with the local landscape. This 'cosmetic' treatment has been widely used for reclaiming tips and it offers the best solution for dealing with the majority of tips. Even if all the bulk fill in new roads were composed entirely of wastes, the amounts used would be insignificant compared with the stockpile of wastes that is potentially available.

The costs of rehabilitating a tip by landscaping and planting can be fairly accurately assessed, so it is possible to quantify the benefit to be gained by using wastes in road construction by taking it to be equivalent to the cost of cosmetic treatment of the tip by landscaping and planting.

Avoidance of borrow pits
As discussed in Parts 1 and 2, aggregates for the construction of the upper layers of a road pavement are almost invariably obtained from commercial quarries and gravel pits. However, for the lower layers, particularly for bulk fill, borrow pits are often used as the source of construction material.

A borrow pit can be sited close to the road project so that haulage distances and the use by construction traffic of public roads are minimized. (It also allows the contractor to use unlicensed vehicles if haulage routes can be so arranged

Fig. 40. Example of a reclaimed colliery spoil tip

to avoid the use of public roads. This point is often quoted in favour of borrow pits but is largely illusory. If the contractor does pay the tax this sum will be added to the contract so that, although the Government gets the benefit of the tax, it pays an equivalent extra sum for the road to be built. The sums involved are, in any case, small).

The advantages to the contractor of using material from a borrow pit have tended to obscure the very real disbenefit that may result from the opening of a pit. Proponents of the use of wastes emphasize the dereliction that a borrow pit can cause, while at the other extreme claims are made that strict enforcement of planning laws would make it impossible for any permanent harmful effects to occur. In fact, although both extremes can occur, the situation lies somewhere in between. Whether or not a borrow pit will permanently disfigure the landscape depends partly on planning regulations and partly on topography. The restoration of borrow pits in undulating countryside can be carried out comparatively easily by landscaping so that, although the topography is affected, the change is not apparent to anybody unacquainted with the area before the operations began. This form of restoration can be carried out quickly and leaves no permanent scars.

In flat country, however, particularly where the water table is high, landscaping may not be a practicable proposition. If the pit is to be reinstated, inert filling materials, which may not be readily available, have to be imported to the site (if they were readily available they could well have been used instead of the borrow material). If the pit is left as a lagoon the land is permanently lost and, although it may be possible to use the lagoon for water sports, the demand for such uses is limited.

Fig. 41. Site of a borrow pit opened for the M4 motorway in 1963 (photographed in 1992)

An examination of the after use of the borrow pits opened up for the eastern sections of the M3 and M4 motorways (Sherwood and Roe 1974) showed that, whereas all the pits opened in undulating country had been effectively restored, many of the pits in the flat areas were lying derelict 15 years after the borrow pits had been opened, a situation that still applies after 30 years (Fig. 41).

Conservation of natural aggregates
The use of waste materials for the construction of those layers where it is possible to use them can have significant environmental benefits and conserve the better quality materials for applications where they are really needed. Conservation of high quality materials in this way will give greater benefit through long-term future use than is foregone by not using them now.

Disadvantages of alternative materials
Increased haulage costs
The use of wastes rather than borrow material usually involves additional haulage costs because a borrow pit will, almost invariably, be closer to the road site than a waste tip (if it is not, the advantage of using waste is so overwhelming as to be self-evident). In some cases it will be necessary to strengthen the roads used for haulage, and the costs of doing so should be added to the extra costs of haulage.

Disturbance caused by haulage
The disturbance caused by haulage is an environmental disbenefit as the haulage lorries use public highways causing congestion, noise, dust and

deposition of material on the road. As this congestion is not easily quantified it tends to be ignored, but it is nonetheless real to people who live near the haulage route. The removal of material from a spoil heap can easily mean the exchange of a permanent, but quiescent, disbenefit, to which local inhabitants have to some extent become accustomed, for a widespread mobile nuisance.

No generally agreed procedure has been developed for quantifying traffic disturbance but, in making a decision on whether wastes should be used, assessments can be made of the length of the haulage route, the volume of traffic generated and the number of people likely to be affected by traffic.

Greater variability of waste materials

All of the waste materials considered in this book have at least some potential for use in road construction as substitutes for naturally-occurring materials, particularly for bulk fill. Most differ in their physical properties and chemical properties to the materials normally used but this does not necessarily place them at a disadvantage.

Particular attention has to be paid to their chemical properties which may be highly variable. Some, notably china clay sand and slate waste, are completely inert chemically and will present fewer problems than are likely to occur with natural materials. Others, such as blast-furnace slag and PFA, although different in chemical composition from the naturally-occurring materials used in road construction, are unlikely to prove to be any more difficult to use. They can present problems that are unique to the particular material but this does not impose any serious limitations on their use.

However, in the case of some waste materials unusual problems arising from their chemical composition do need to be considered. This does not mean that all such materials should be regarded with suspicion but it does mean that tests that would not normally be considered necessary have to be carried out to ensure that they are suitable for use.

In considering the use of wastes, care should therefore be taken not to regard them as being equivalent in all respects to the natural materials they are expected to replace. (The recent review for the Department of the Environment, (Whitbread *et al*. 1991) of the occurrence of waste materials can be criticized for making this assumption). This is rarely the case. Aggregates produced by commercial suppliers are processed to meet the requirements of a particular specification. The user can reject the material if it fails to meet the specification requirements and can require the supplier to rectify any problem that occurs. Although waste materials can be processed in the same way as natural aggregates this greatly increases the costs and cannot be justified unless large quantities are being regularly taken from the stockpile. This means that they require more testing on site than do commercially-produced aggregates.

Possibility of long-term pollution problems

Naturally-occurring materials used in road construction are most unlikely to give rise to any problems of pollution arising from the leaching of soluble materials into water-courses. Most are chemically inert and contain insignificant amounts of soluble material that could be removed by leaching. The very fact that they occur naturally gives confidence in their use because any soluble materials that they might contain would have given rise to problems in the areas in which they occur. If such problems had arisen, this fact would have been realized so that their use could be avoided.

This is also true of china clay wastes and slate waste which are naturally-occurring quarried materials. It is not true, however, of most of the other materials considered in this book. Many of these contain soluble materials that could be a potential problem, and this fact is considered later.

Conclusions

The preceding sections showed that the use of alternative materials involves a benefit (A) usually as a result of removing a waste tip, but also involves a disbenefit made up of transport costs (C), the disturbance caused by haulage traffic (D), the extra costs (E) arising from the greater variability of waste materials and a factor (F) arising from the greater potential for water pollution if waste materials are used. Therefore, as it is a fundamental concept that the money spent on controlling a disbenefit should not exceed the cost that the disbenefit is causing, the relation between A, C, D, E and F can be written in the form

$$A \geq C + D + E + F$$

However, if waste material is not used the material has to be obtained from a borrow pit which involves a disbenefit (B). Therefore the above equation has to be modified to

$$A + B \geq C + D + E + F$$

This relation is deceptively simple and if all the factors could be measured in the same units there would be no problem in deciding when to use wastes and when to use borrow pits. As the above discussion has shown, only A, C and, possibly, E of the factors to consider can be measured in monetary terms; even if an assessment is made of B and D the results are not in the same units. How, for example, can the disbenefits caused by borrow pits, which may be permanent, be compared with the temporary disturbance caused by traffic, particularly as the people suffering from the effects of one may not be the same group of people affected by the other.

The answer is that somebody has to make a subjective judgement, but in doing so all the factors have to be given due consideration. As the decision is subjective it will not generally prove possible to convince everybody that the correct decision has been reached. For this reason it is necessary to consider, in the absence of a generally accepted method of comparing all the factors involved, what further steps might be taken to improve the present position. Ways in which this might be done are therefore considered in the next chapter.

13. Encouraging the use of alternative materials

The first chapters in Part 3 have shown that there are many environmental benefits to be gained by using alternative materials as substitutes for the materials that are traditionally used in road construction. However, if alternatives are to be so much as considered as potential substitutes, positive steps need to be taken to encourage their use. Possible methods by which this can be done are considered in this chapter.

Amendments to existing standards and specifications

The situation with regard to standards for alternative materials was considered in Part 1, which discussed the European standards that are being prepared to cover the use of alternative materials. Good progress is being made with the preparation of these standards and, in the long term, the relative absence of standards that give the requirements for alternative materials will be rectified.

In the meantime, the Specification for Highway Works (1991a) gives valuable guidance on the use of many of the alternatives considered here. For example, it mentions by name many of the materials as being suitable for particular applications. This is true of blast-furnace slag and crushed concrete which are named as being suitable for nearly all the uses covered by this book. Where they are considered suitable, other materials such as colliery spoil and PFA are also mentioned by name but some materials, notably china clay sand and slate waste, are not mentioned at all. Absence of a specific mention does not, however, preclude use if a material fulfilled all the requirements for a particular use and it could be argued that it fell into a category that was permitted. For example, slate waste is a type of crushed rock, china clay sand differs little from quarried sands, and spent oil shale is very similar to burnt colliery spoil.

It can therefore be concluded that the specification takes a permissive attitude to the use of wastes and, generally speaking, it would be difficult to argue that it is unduly restrictive. There are certain anomalies in the specification relating to restrictions on particular materials which seem difficult to justify; these are considered in the relevant sections of the report. However, these anomalies are relatively minor and even if rectified would do little to add to the amount of waste materials already being used in road construction.

This conclusion is borne out by two critical appraisals for the Department of the Environment of the specifictions for road and building construction in use in this country (Collins et al. 1993, Collins and Sherwood 1995). These confirmed that the specification allowed a wide range of materials to be used in road construction. They also found that although all highway authorities accept the specification as a basis for their own specifications, many make significant departures from it. These departures were, on the whole, likely to lead to more stringent requirements being imposed rather than to any relaxations.

Assessment of experience in other countries
OECD review

OECD (1977) reviewed the use of waste materials and by-products in road construction in member countries — Belgium, Canada, Denmark, Finland, France, Germany, Italy, Netherlands, Spain, Switzerland, the UK and the USA took part. This review showed that most of the larger countries produced considerable quantities of colliery spoil, PFA, ferrous slags and demolition wastes, and all the other materials discussed in this book were produced in at least one other country. Some wastes discussed in the OECD report did not occur in the UK, while in the case of china clay sand the UK and the USA were the only producers.

For each waste and by-product judged by the OECD committee to have potential for use in road construction, a detailed discussion of the physical and chemical properties, established uses and the problems that might occur was given in the report. The review showed that, at that time, the UK was well to the fore in developing uses for waste materials and by-products in road construction and, for the most part, had more to offer by way of experience in their use than it could learn from experience in other countries.

Few of the materials available in the UK were being used in other countries for purposes for which those materials were not already being used or considered for use in this country. The major exception was the use of blast-furnace slag, particularly in France where, relative to air-cooled slag, granulated slag was produced in much larger quantities and found many outlets in road construction (as discussed in the chapter on blast-furnace slag in Part 2).

SPRINT review

The European Community has a Strategic Programme for Innovation and Technology transfer (SPRINT) in Road Construction. This aims to stimulate technology transfer and innovation among the member countries. At the second SPRINT meeting in Rotterdam in June 1992 the programme concentrated on the use of colliery spoil, metallurgical slags, power station wastes, construction and demolition wastes and incinerator ash. The papers given at this meeting did not reveal any uses of alternative materials not considered here, but they did show that since the OECD review the UK had lagged somewhat in the use of such materials and within Europe there was a greater emphasis on the environmental benefits of using them.

The UK, France and Germany produced by far the greatest amounts of colliery spoil; it was claimed (Dac Chi 1992) that, whereas in the UK only $1 \cdot 1\%$ of the current production of colliery spoil was used, in Germany and France the corresponding figures were 21% and 170%, respectively. The French figure was based not only on current production but took into account the use of material from existing tips. However, even if allowance is made for this, it is clear that greater use is now being made of alternative materials in many European countries than is the case in the UK. Some of the reasons for this, and of how the use of alternatives could be encouraged, are discussed later.

Guidance on the use of alternative materials

One factor that could inhibit the greater use of alternative materials in road construction is the absence of guidance on their use. The Specification for Highway Works (1991a), together with the accompanying Notes for Guidance and the advice notes which are issued from time to time, advise on the construction methods to be employed with all roadmaking materials and give some guidance on the use of certain wastes. There is also BS 6543 (1985a) on the use of wastes in civil engineering, but it does not seem to be widely known (Collins *et al*. 1993). Apart from these publications, producers of the major wastes publish technical information which is freely available, and symposia and conferences are held on the use of wastes.

Guidance on the use of particular wastes is not therefore lacking for those who wish to use them. More important is the attitude of prospective users of wastes. If they are not interested or have no incentive to use wastes, no matter how much information is published it will remain unread.

Changing attitudes to the use of alternative materials

The survey by Collins *et al*. (1993) did not reveal any widespread resistance to the use of alternative materials; they were not perceived to be of lower grade, or more troublesome to use. The main reason for not using them was that there was rarely any incentive to do so. The naturally-occurring materials used in road construction are relatively cheap and readily available (perhaps too much so) and their cost forms only a small proportion of the total cost of the road. There is thus a tendency to 'play-safe', even if such materials do cost more; criticism of needless expenditure is muted compared with the criticism that occurs if a costly failure should arise through the use of a material that subsequently proves to be unsuitable.

Any environmental benefits to be gained by using lower-quality, but acceptable, materials are ignored because there is no provision for taking them into account. For this reason those responsible for the choice of material are likely to be influenced by a relevant maxim from another field that 'nobody was ever sacked for choosing IBM'. The use of waste materials will thus remain a laudable aim which will not be fulfilled unless some means can be found of swinging the balance in favour of their use. How this might be done is considered next.

Administrative measures to increase the use of alternatives

In the absence of any generally accepted method of putting a monetary value on the environmental benefits to be gained from the use of wastes, other methods have to be used to encourage their use. Policy methods that have been put forward are discussed in the remainder of this section.

Nomination of the use of a particular source of waste material

The use of alternative materials rather than naturally-occurring materials is such an attractive proposition that there has been much pressure on the Department of Transport to utilize them in its road building programme. As can be seen from the frequent references to waste materials in the Specification for Highway Works (1991a), the DOT is favourably disposed to the use of wastes. However, it has always been reluctant to insist that alternatives such as waste materials should be used, even when they are readily available.

This stems from the fear that if it did so and the nominated source gave rise to unforeseen problems (either real or imaginary), it would be exposed to claims (possibly unreasonable) for compensation by the contractor. It has therefore preferred to specify a wide range of materials, which in many instances includes waste materials and by-products, for any given purpose. It then allows contractors complete freedom of choice of the material (provided it complies with the Department's specification) within this range. The Department's attitude on this subject has been widely criticized, not only by environmental pressure groups but also by official bodies such as the Advisory Committee on Aggregates (1976) and the Royal Commission on Environmental Pollution (1974).

As a result of these pressures the Department reviews its policies from time to time. In the latest reviews (Department of Transport 1986b, DOE/DOT/WO 1987) the Department maintains its stance that tenderers should be left to choose their source of fill purely on a commercial basis. The review points out that the demand for imported fill for road schemes is small compared with the available sources of waste materials and that the contribution that road building can make to the problem of dealing with waste is limited. It endorses the view, expressed by the Committee on Planning Control over Mineral Working (1976), that in planning new roads, highway authorities should discuss with the planning authority at an early stage whether there are likely to be environmental constraints in the use of locally excavated fill material.

Imposition of a tax on the use of naturally-occurring materials

A recent report on the occurrence and utilization of waste materials (Whitbread *et al.* 1991) concluded that the most promising line of policy intervention would be to introduce a tax on the use of primary aggregates. Such a price rise for primary aggregates could have two significant impacts on the economic benefits of a given road proposal:

(*a*)   it could promote more careful scrutiny of the net economic benefits of the road proposal

(*b*)   it would encourage greater efficiency in use of primary aggregates and greater use of alternative materials.

Whitbread *et al.* (1991) considered that a general charge of 15% on the price of primary aggregates could bring about a switch of 10 million tonnes per year from primary to secondary materials. According to their report this approach has already been adopted in Denmark which has introduced a tax of about £0·25/tonne on primary mineral extraction and a charge of about £12/tonne on all landfill waste disposal. This is said to have had a dramatic effect in altering industry practice with regard, in particular, to demolition and construction waste disposal and the amount of recycling.

The recommendation has been criticized by BACMI (1991) which claims that the tax on primary aggregates could increase construction costs by £1·25 billion by 2011 to achieve, in the view of BACMI, an additional increase in usage of waste materials of 7 million tonnes.

Imposition of a tax on the disposal of construction wastes to landfill

This is a mirror image of the tax on the use of primary aggregates considered above. At present much material that has potential for use in construction work is simply dumped as landfill where, for example, it is used to reclaim land that has been used for aggregate excavations. Although the use of such material is not without benefit in land reclamation, it would make more sense to use it rather than to create more holes in the ground to obtain primary aggregate. Such a tax would therefore encourage the use of alternative materials. The Government has recently announced that it intends to impose such a tax with effect from 1996.

Imposition of a requirement that all projects should incorporate a proportion of waste or recycled material.

A less controversial approach than that of a tax on naturally-occurring materials and which does not have the disadvantage of the above nomination approach is to impose a general requirement that a given proportion of the materials used in a construction project should be either waste or recycled material. This is the approach that has been adopted in some countries, e.g. Germany and the Netherlands. It is backed-up in Germany by a law which prohibits the disposal of mineral wastes which are capable of being recycled.

Such requirements have aroused interest in the use of waste and recycled materials and may well mean that more of them are used than the law demands. This also means that standards and specifications are produced for their use that mirror those already in existence for naturally-occurring materials. Sweere (1989) reported that in the Netherlands half of the market for granular sub-base materials is made up of recycled construction and demolition waste. The granular materials are produced in special plants in a process of crushing and cleaning. The cleaning involves the magnetic removal of iron and the removal of wood, plastics, etc. through a washing process. Sweere (1991) also made reference to Dutch standards that give the requirements for recycled materials.

Reservation of high quality materials for high-grade use

This is a mirror image of the imposition of a requirement that all projects should incorporate alternative materials. It would reserve higher-grade materials for those purposes for which the scarce properties of the material were essential. Thus, valley gravels which could be washed and used as high-grade aggregates would be used for ready-mixed concrete where alternative materials are not readily available, but never in an unbound form where numerous alternatives exist. However, such a proposal would involve some form of end-use control over quarried materials and, apart from being unpopular with producers, would be difficult to enforce.

14. Policy and controls on the supply and use of construction materials

The enormous amounts of materials (principally aggregates) and the effects that the supply and use of these may have on the environment inevitably mean that the Government has been forced to play an active role in ensuring that the naturally-occurring materials required by the construction industry are made available. At the same time the amounts of waste materials that are produced and the fact that some of these may be hazardous has also meant that Government action has had to be taken on the use and disposal of these products. Earlier parts of this book have shown that some of these waste materials can be used as alternatives to naturally-occurring materials. Policy and control on the supply and use of traditional and alternatives therefore become interlinked. Government policy on both of these issues was summarized in a paper (Department of the Environment 1994a) which states:

> With increasing demands for minerals the key issues for sustainability in this sector are:
>
> (a) to encourage prudent stewardship of mineral resources while monitoring necessary supplies
> (b) to reduce environmental impacts of minerals provision both during minerals extraction and when restoration has been achieved.
>
> The Government will therefore:
>
> (a) continue research on the availability of mineral resources and the environmental costs and benefits of using different sources
> (b) promote more efficient use of mineral resources in general, for example, encouraging recycling of materials and substitution of alternative materials, where appropriate.

In this chapter the main policy and control factors affecting both naturally-occurring and alternative materials are discussed. The aim is to give a background to official policy with regard to the controls that are imposed on the use of construction materials; it is not intended to be exhaustive as some aspects are complex and are further complicated by frequent additions and amendments.

Policy on the availability of construction materials

It is Government policy to ensure that there is an adequate supply of minerals to meet the needs of the construction industry. To achieve this aim Regional Aggregates Working Parties (RAWPs) have been set up to prepare supply-and-demand forecasts for their particular region using data made available by the DOE of long-term forecasts for primary aggregates. The regional commentaries prepared by each RAWP form part of the Mineral Planning Guidance (MPG) Notes issued by the DOE. These guidance notes set the parameters for minerals planning policy at national and regional level. The latest edition of MPG 6 (DoE 1994b) — Minerals Planning Guidelines for Aggregates Provision in England — states that one of the aims of the Guidance Note is to:

> provide guidance on how an adequate and steady supply of material to the construction industry, at a national, regional and local level, may be maintained at the best balance of social, environmental and economic cost, through full consideration of all resources and the principles of sustainable development.

The 1994 edition of MPG 6 places greater emphasis on sustainability than hitherto. It identifies the potential conflict between increasing aggregate demand and the need to limit the impact of extraction industries on the natural environment. The Guide notes the importance of the use of secondary aggregates and recycled materials which it estimated accounted for 10% of the market in 1989 and set targets of 40 million tonnes per annum by 2001 and 55 million tonnes by 2006 (in 1989, 40 million tonnes would have represented 17% of the market).

Six months after the publication of MPG 6, the Royal Commission on Environmental Pollution published a report (1994), containing the recommendation that the proportion of recycled materials used in road construction should be doubled by 2005 and doubled again by 2015. The targets set by MPG 6 and the Royal Commission are not incompatible, but the realization of these targets would require planning policy changes to encourage alternative sources of construction materials to enter the market — how this might be done has already been considered in Section 13.

Legislation concerning construction materials
Natural materials

The excavation of aggregates and fill materials is subject to planning controls exercised by the local planning authority (LPA). An LPA can refuse permission for excavation of aggregates but its ability to do so is limited and the applicant can always appeal to the DOE against the refusal. LPAs are required to prepare a local Minerals Plan which must incorporate the supply and demand figures prepared by the RAWPs. LPAs cannot challenge the national forecasts, only the amount of minerals they are expected to produce towards meeting total forecast demand.

In the case of borrow pits opened up to provide fill materials for road construction the LPA has more scope because, generally speaking, the material from a borrow pit will not be classed as an aggregate. The arguments for and against the use of naturally-occurring materials from borrow pits rather than the use of waste materials was discussed earlier in Part 3. If planning consent is given by the LPA for the opening of a borrow pit it will invariably be subject to conditions requiring that the pit should be restored to some beneficial use.

As the opening of a borrow pit to obtain fill material is subject to its control, the LPA can to some extent (but not wholly, because there is a right of appeal to the DOE if a planning application is refused) play a part in deciding on the source and type of material that should be used for bulk fill. If suitable waste materials are available within a reasonable distance of the site the LPA would probably insist on their use and refuse permission for a borrow pit to be opened.

Waste materials

Controlled wastes. The disposal of waste materials is covered by the Environmental Protection Act 1990 (DOE 1992). Not all parts of this Act are in force, but its provisions are being introduced progressively by Statutory Instruments. 'Controlled wastes', as defined by the Act, means household, industrial and commercial waste. Briefly, under the Act a material is defined as a waste if the answer to any of the following questions is 'yes'.

(*a*)   Is it what would ordinarily be described as a waste?
(*b*)   Is it a scrap material?
(*c*)   Is it an effluent or other unwanted surplus substance?
(*d*)   Does it require to be disposed of as broken, worn out, contaminated or otherwise spoiled?
(*e*)   Is it being discarded or dealt with as if it were a waste?

Waste from commerce or industry which falls into any of these categories is a 'controlled waste' and the producer of the waste has 'a duty of care' to ensure that its disposal does not cause any environmental problems. To ensure that it does not, if the waste leaves the premises on which it is produced the waste carrier must be registered with a waste regulation authority and transfer documentation will be required. Wastes to be treated as industrial wastes include:

(*a*)   waste arising from tunnelling or from any excavation
(*b*)   waste removed from land on which it has been previously deposited and any soil with which such wastes have been in contact.

At first sight it would seem, therefore, that all the materials discussed in this book fall within the controlled wastes covered by the Act. However, wastes from mining and quarrying operations are specifically excluded. The argument in favour of their exclusion is based on the fact that disposal in the mine or quarry falls within the terms of the original planning consent. This means that the Act does not apply to colliery spoil, china clay sand, slate waste and spent oil shale (see below). Nor does it apply to the current production of blast furnace slag as it is produced as a manufactured product all of which is sold to the construction industry.

Hence only construction and demolition wastes, FBA and PFA fall into the category of controlled waste. But stockpiles of construction and demolition wastes are also exempted from the requirement to obtain a disposal licence 'provided that the deposit is made for the purposes of construction currently being undertaken on the land on which the waste is deposited (or will be used for future construction within three months)'. If any doubts exist about the classification of a particular material, the local Waste Management Authority should always be consulted.

Other wastes. The fact that mining and quarry wastes are not classified as controlled wastes does not mean that there are no controls placed on them. Producers must apply for planning permission to the relevant minerals planning authority for all new spoil disposal schemes and, if approved, conditions are attached to the planning permission to safeguard the surrounding environment. However, some tipping still takes place on tips of long standing with no conditions attached, as permitted development under the General Development Order. The responsibility for enforcing the planning regulations rests with the LPA which, outside the metropolitan conurbations of England and Wales, is the appropriate County Council.

The safety of spoil tips is covered by legislation designed to control surface tipping rather than to eliminate it. Thus the Mines and Quarries (Tips) Act of 1969 was passed as a response to the Aberfan disaster of 1966 and was designed to ensure that spoil tips do not constitute a hazard to life and property. This Act has meant that the conical-shaped spoil tips of the past have largely disappeared to be replaced by tips with plateau-shaped surfaces (Fig. 12) with slopes that are more stable and less likely to slip. Precautions are also taken to ensure that tips do not catch fire, with the result that the amount of burnt colliery spoil that is available is steadily declining.

Recycled materials. Recycled materials can fall under the heading of controlled wastes but there is the added complication that recycling plants also need to have planning consent. A fixed site recycling plant involving the import of materials and the export of an improved recycled product is an unwelcome neighbour. A planning consent may therefore be granted only after consideration of factors relating to both the site and the proposed operation. The final consent may be conditional and impose requirements to limit the impact of noise, visual intrusion and air pollution.

Health and safety considerations
Hazards to site personnel
None of the materials considered in this report is particularly harmful and, on the whole, their extraction and use requires no additional precautions over and above those normally required in civil engineering construction, which are covered by the Health and Safety at Work Act 1974 and earlier legislation. However, certain problems arise which would not normally be encountered with natural materials. The combustion of colliery spoil can give rise to high local levels of toxic gases such as carbon monoxide, sulphur dioxide and hydrogen sulphide which may be a hazard to workmen removing spoil where excavation exposes an area of burning material. The instability of seemingly hard surfaces of waste deposits may also be a problem. Lagoons of PFA may form a surface crust that appears to be quite stable, and undergound fires in colliery spoil tips may result in the formation of similar crusts in spoil deposits.

Pollution of ground water

With the exceptions of china clay sand and slate waste, many of the other materials considered in this book may contain traces of toxic compounds that, given the right conditions, could leach out and pollute water courses. The possibility of this depends on the concentration of the toxic substance, the quantity of material being used and the readiness with which it can be brought into solution.

Materials bound with bitumen or cement are not likely to present any problems for two reasons. First, the particles are encapsulated by a matrix of bitumen or cement which impedes, even if it does not completely prevent, the passage of water into the individual particles. Second, bound materials are mainly used in the upper layers of the road structure in layers that are thin compared with the thicknesses of the underlying layers (see Figs 6 and 7). Apart from the fact that the rate of leaching is likely to be low, the total amount of toxic material that could be leached out is also likely to be low. A similar argument can be applied to unbound layers provided that they are not used in extensive thicknesses and they are not in a permanently wet condition.

However, the lower layers of the road structure may contain considerable volumes of material. This is particularly true of the volume of material in road embankments which cross low-lying areas where high embankments may be required and the material is in a permanently saturated condition and where drainage is poor. Due to the large volumes of embankment material that may be used the amount of toxic material present may become significant even if its concentration is low. This is generally a localized problem because the material is likely to be at least partially encapsulated by materials of very low permeability and the rate of leaching is slow. Moreover, the subsequent leachate is diluted by heavy rainfall so that any toxic effect of the leached compounds is reduced to less than the threshold for toxicity.

Problems could arise, however, where the drainage from the road embankments discharges directly to rivers as pollution of the water may seriously affect aquatic life. Further problems may arise if the rivers receiving discharges are sources of public water supply, or where embankments are constructed close to springs or ground water sources similarly used for the public supply. Although in such circumstances the probability of toxic effects from leached materials would be low, leached materials may give rise to complaints of taste and odour of the water.

Experience is that the use of unburnt colliery spoil and most quarry wastes are not likely to cause problems of water pollution. Other types of wastes, such as old slag deposits, may lead to problems, however, particularly where there are traces of toxic compounds such as those of cadmium and mercury. If doubt exists about water pollution arising from the use of substantial amounts of waste, consultations should be held with the appropriate water authorities at the planning stage to make sure that public water supplies are protected and that the materials to be used do not constitute a risk to the local water courses. This may require measures to reduce risks of pollution to an acceptable level.

In constructing an embankment from wastes or by-products the aim should be to achieve run-off of rain rather than percolation through the fill. Adequate compaction before cladding with topsoil and the establishment of vegetation are essential.

Pollution of the atmosphere

Problems arising from air pollution from burning colliery spoil have already been considered. The only other type of major air pollution which can arise is from dust, which can be troublesome in windy weather, particularly with small particles of light material such as PFA. It is therefore important that materials should be transported in sheeted lorries and be laid and compacted on arrival on site. Residual problems can be coped with by spraying the compacted material with water and keeping traffic off the surface.

PART 4
REFERENCES

AASHTO M 147-65 (1990). *Standard specification for materials for aggregate and soil—aggregate sub-base, base and surface courses*. American Association of State Highway and Transportation Officials, Washington, USA.

ADAMS J. (1991). *Determined to dig — the role of aggregates demand forecasting in national minerals planning guidance*. Council for the Protection of Rural England, London.

ADVISORY COMMITTEE ON AGGREGATES (1976). *Aggregates: the way ahead*. HMSO, London.

ANNUAL ABSTRACT OF STATISTICS (1991). HMSO, London.

ASTM C595-76 (1976). *Standard specification for blended hydraulic cements*. American Society for Testing Materials, Philadelphia.

ASTM D2940-74 (1985). *Standard specification for graded aggregate material for bases or sub-bases for highways or airports*. American Society for Testing Materials, Philadelphia.

BACMI (1991). *The occurrence and utilisation of mineral and construction wastes: A BACMI response to the report by Arup Economics Commissioned by DOE*. BACMI, London.

BAMFORTH P.B. (1992). (DHIR R.K. and JONES M.R. (eds)). What PFA does to concrete. *National seminar on the use of PFA in construction*, Dundee, 145—156.

BRITISH GEOLOGICAL SURVEY (1992). *United Kingdom Minerals Yearbook 1991*. British Geological Survey, Keyworth.

BS 146 (1973). *Specification for Portland—Blastfurnace cement*. British Standards Institution, London.

BS 1047 (1983a). *Air-cooled blastfurnace slag aggregate for use in construction*. British Standards Institution, London.

BS 3892 Part 1 (1983b). *Specification for pulverised fuel ash for use as a cementitious component of structural concrete*. British Standards Institution, London.

BS 3892 Part 2 (1984). *Specification for pulverised fuel ash for use in grouting and for miscellaneous uses in concrete*. British Standards Institution, London.

BS 6543 (1985a). *Guide to the use of industrial by-products and waste materials in building and civil engineering*. British Standards Institution, London.

BS 6588 (1985b). *Specification for Portland pulverised fuel ash cement*. British Standards Institution, London.

BS 6610 (1985c). *Specification for pozzolanic cement with pulverised fuel ash as a pozzolana*. British Standards Institution, London.

BS 6699 (1986). *Specification for ground granulated blastfurnace slag for use with Portland cement*. British Standards Institution, London.

BS 812: Part 118 (1988). *Testing aggregates: methods for the determination of sulphate content*. British Standards Institution, London.

BS 812: Part 121 (1989). *Testing aggregates: methods for the determination of soundness.* British Standards Institution, London.

BS 812: Part 124 (1989). *Testing aggregates: method for the determination of frost-heave.* British Standards Institution, London.

BS 1377: Part 3 (1990) *Methods of tests for soils for civil engineering purposes: chemical tests.* British Standards Institution, London.

BS 882 (1992). *Specification for aggregates from natural sources for concrete.* British Standards Institution, London.

BULLAS J.C. and WEST G. (1991). *Specifying clean hard and durable aggregate for bitumen macadam roadbase.* Transport and Road Research Laboratory, Crowthorne, UK, Research Report LR 284.

BURNS J. (1978). *The use of waste and low-grade materials in road construction — 6. Spent oil shale.* Transport and Road Research Laboratory, Crowthorne, UK, Report LR 818.

CARR C.E. and WITHERS N.J. (1987). (RAINBOW A.K.M (ed.).) The wetting expansion of cement-stabilized minestone — an investigation of the causes and ways of reducing the problem. *Reclamation, treatment and utilization of coal mining wastes.* Elsevier, Amsterdam. 545–559.

CEGB (1972). *PFA utilization.* Central Electricity Generating Board, London.

CEN (1993). *Specification for fly ash for hydraulic bound mixtures (non-factory blended binders).* Draft European Standard, CEN Technical Committee 227, Document N112E.

CEN (1994). *Tests for chemical properties of aggregates. Part 1: chemical analysis.* Draft European Standard, CEN Technical Committee TC 154, Document EN1744-1.

CIPFA (1988). *Waste disposal statistics 1988–89 estimates.* Chartered Institute of Public Finance and Accountancy, London.

CLARKE B. (1992). (DHIR R.K. and JONES M.R., (eds)) Structural fill. *National seminar on the use of PFA in construction*, Dundee, 21–32.

COLLINS R.J. (1990). Case studies of floor heave due to microbiological activity in pyritic shales. *Proc. Conf. on Microbiology in Civil Engineering.* E. & F.N. Spon, London, 288–295.

COLLINS R.J. (1993). Personal communication.

COLLINS R.J. and SHERWOOD P.T. (1994). Use of wastes and recycled materials as aggregates: standards and specifications. (To be published by BRE in 1995).

COLLINS R.J., SHERWOOD P.T. and RUSSELL A.D. (1993). *Efficient use of aggregates and bulk construction materials. Volume 1: an overview. Volume 2: technical data and results of surveys.* Building Research Establishment, Garston, UK, Reports BR243 and BR244.

COMMISSION ON ENERGY AND THE ENVIRONMENT (1981). *Coal and the environment.* HMSO, London.

COMMITTEE ON PLANNING CONTROL OVER MINERAL WORKINGS (1976). *Final report.* HMSO, London.

CORNELIUS P.D.M. and EDWARDS A.C. (1991). *Assessment of the performance of off-site recycled bituminous material.* Transport and Road Research Laboratory, Crowthorne, UK, Research Report LR 305.

CPRE (1993). *Driven to dig — road building and aggregates demand.* Council for the Protection of Rural England, London.

CROCKETT R.N. (1975). *Slate — mineral dossier No. 12.* Mineral Resources Consultative Committee, HMSO, London.

DAC CHI N. (1992). Use of carbonaceous shales and metallurgical slags in road techniques. *SPRINT Workshop on Alternative Materials in Road Construction*, Rotterdam.

DAWSON A.R. and BULLEN D. (1991). Furnace bottom ash: its engineering properties and its use as a sub-base material. *Proc. Instn Civ. Engrs.* 1991, Part 1, **90**, 993–1009.

DEPARTMENT OF THE ENVIRONMENT (1991a). *Survey of derelict land in England 1988.* HMSO, London.

DEPARTMENT OF THE ENVIRONMENT (1991b). *Environmental effects of surface mineral workings.* HMSO, London.

DEPARTMENT OF THE ENVIRONMENT (1992). *The Environmental Protection Act 1990 — parts II and IV: the controlled waste regulations 1992.* DOE Circular 14/92. HMSO, London.

DEPARTMENT OF THE ENVIRONMENT (1994a). *Sustainable development — The UK strategy. Summary report.* DOE, London.

DEPARTMENT OF THE ENVIRONMENT (1994b). *Guidelines for aggregates provision in England, minerals planning guidance note MPG 6.* HMSO, London.

DOE/DOT (1992). *Joint Memorandum by the Departments of the Environment and Transport to the Royal Commission on Environmental Pollution Transport and Environment Study.*

DOE/DOT/WO (1987). *Use of waste material for road fill.* DOE Circular 20/87, DOT Circular 3/87, WO Circular 36/87. Joint Circular by the Departments of the Environment and Transport and the Welsh Office. HMSO, London.

DEPARTMENT OF TRANSPORT (1986a). *Specification for Highway Works.* HMSO, London, 6th edn.

DEPARTMENT OF TRANSPORT (1986b). *Report of the Interdepartmental Committee on the use of waste materials for road fill.* HMSO, London.

DEPARTMENT OF TRANSPORT (1987). Structural design and new road pavements. Department of Transport Highways and Traffic Advice Note HA 35/87 and Departmental Standard HD 14/87.

DEPARTMENT OF TRANSPORT (1991a). *Specification for Highway Works 1991. Volume 1: manual of contract — Documents for Highway Works. Volume 2: manual of contract — Documents for Highway Works.* Notes for Guidance. HMSO, London.

DEPARTMENT OF TRANSPORT (1991b). Earthworks — design and preparation of contract documents. Department of Transport, London, Advice Note HA 44/91.

DUNSTAN M.R.H. (1981). *Rolled concrete for dams — a laboratory study of the properties of high fly-ash concrete.* Construction Industry Research and Information Association, London, Technical Note 105.

FRANKLIN R.E., GIBBS W. and SHERWOOD P.T. (1982). *The use of pulverised fuel ash in lean concrete. Part 1 — laboratory studies.* Transport and Road Research Laboratory, Crowthorne, UK, Supplementary Report SR 736.

FRASER C.K. and LAKE J.R. (1967). *A laboratory investigation of the physical and chemical properties of burnt colliery shale.* Road Research Laboratory, Crowthorne, UK, Report LR 125.

GASPAR L. (1976), Les cendres volants et le laitier granulé en construction routière. *Bulletin de Liaison*, **86**, 135–143.

GOULDEN E.R. (1992). Slate waste aggregate for unbound sub-base layers. University of Nottingham, MSc thesis.

GUTT W., NIXON P.H., SMITH M.A., HARRISON W.H. and RUSSELL A.D. (1974). *A survey of the location, disposal and prospective uses of the major industrial by-products and waste materials.* Building Research Establishment, Garston, UK, Current Paper CP 19/74.

HARDING H.M. and POTTER J.F. (1985). *The use of pulverised fuel ash in lean concrete. Part 2 — pilot-scale trials.* Transport and Road Research Laboratory, Crowthorne, UK, Supplementary Report SR 838.

HAWKINS A.B. and PINCHES G.M. (1987). Cause and significance of heave at Llandough Hospital, Cardiff — a case history of ground floor heave due to gypsum growth. *Quarterly Journal of Engineering Geology*, **20**, 41–57.

HOCKING R.N. (1994). China clay wastes. *Seminar on the use and improvement of marginal and waste materials*. Geological Society, London.

HOSKING J.R. and TUBEY L.W. (1969). *Research on low-grade and unsound aggregates*. Road Research Laboratory, Crowthorne, UK, Report LR 293.

HOWARD HUMPHREYS and PARTNERS (1994). *Managing demolition and construction wastes*. Report of the study on the recycling of construction and demolition wastes in the UK. HMSO, London.

HUBERT P.A. (1987). Energy use comparison between conventional reconstruction and hot drum mix recycling for major trunk roads. *Proc. seminar highway construction and maintenance*. PTRC, London.

KENT COUNTY COUNCIL (1985). Road trial of phosphoric slag as roadbase. (Unpublished report prepared for Civil and Marine Ltd).

KETTLE R.J. and WILLIAMS R.I.T. (1978). Colliery shale as a construction material. *International conference on the use of by-products and waste in civil engineering*. Paris.

LEA F.M. (1970). *The chemistry of cement and concrete*. Edward Arnold, London, 3rd edn.

LEE A.R. (1974). *Blastfurnace and steel slag*. Edward Arnold, London.

MEARS (1975). Mears use slate waste as sub-base on North Wales road contract. Mears Construction Ltd, Press Release.

MERCER J. and POTTER J.F. (1990). Recycling bituminous roads: research and implementation. *BACMI seminar: The new road programme: Blacktop recycling — policy and practice*. BACMI, London.

MINISTERE DES TRANSPORTS (1980). *La technique Française des assises de chaussées traitees aux liants hydraulique et pozzonaniques*. Minstère des Transports, Direction des Routes et de la Circulation routière, Paris (English translation).

MULHERON M. (1991). Recycled demolition waste. *Unbound Aggregates in Construction*, Nottingham.

MULHERON M. and O'MAHONEY M.M. (1990). Properties and performance of recycled aggregates. *Highways and Transportation*, **2**, No. 37.

NATIONAL COAL BOARD (1983). *Cement bound minestone — user's guide for pavement construction*. NCB Minestone Executive, Whitburn, UK.

NATIONAL POWER (1990). *National Ash products and services*. National Power, Swindon.

NEW CIVIL ENGINEER (1980). Sad Canterbury tale follows minestone's early success. *New Civil Engineer*. 27 November.

NITRR (1986). *Cementitious stabilizers in road construction*. National Institute for Transport and Road Research, CSIR, South Africa, TRH 13.

NIXON P.J. (1978) Floor heave in buildings due to the use of pyritic shales as fill material. *Chemistry and Industry*, 4 March, 160–164.

OECD (1977). *Use of waste materials and by-products in road construction*. Organisation for Economic Co-Operation and Development, Paris.

OECD (1981). Utilisation des granulats margineaux en construction routières (unpublished).

OECD (1984). *Road binders and energy savings*. Organisation for Economic Co-Operation and Development, Paris.

O'MAHONEY M.M. (1990). Recycling of materials in civil engineering. University of Oxford, PhD thesis.

PLEASE A. and PIKE D.C. (1968). *The demand for road aggregates*. Transport and Road Research Laboratory, Crowthorne, UK, Report LR 185.

PORTLAND CEMENT ASSOCIATION (1971). *Soil—cement laboratory handbook*. Portland Cement Association, Skokie, Illinois.

PORTLAND CEMENT ASSOCIATION (1977). *Suggested specifications for soil—cement waste course*. Portland Cement Association, Skokie, Illinois.

PORTLAND CEMENT ASSOCIATION (1979). *Soil—cement construction handbook*. Portland Cement Association, Skokie, Illinois.

POTTER J.F., SHERWOOD P.T. and O'CONNER M.G.D. (1985). *The use of pulverised fuel ash in lean concrete. Part 3 — field studies*. Transport and Road Research Laboratory, Crowthorne, UK, Supplementary Report SR 842.

RAINBOW A.K.M. (1989). *Geotechnical properties of United Kingdom minestone*. British Coal, London.

ROE P.G. (1976). *The use of waste and low-grade materials in road construction — 4. Incinerated refuse*. Transport and Road Research Laboratory, Crowthorne, UK, Report LR 728.

ROYAL COMMISSION ON ENVIRONMENTAL POLLUTION (1974). *Fourth report — pollution control: progress and problems*. HMSO, London.

ROYAL COMMISSION ON ENVIRONMENTAL POLLUTION (1985). *Eleventh report — managing waste: the duty of care*. HMSO, London.

ROYAL COMMISSION ON ENVIRONMENTAL POLLUTION (1993). *Seventeenth report — incineration of waste*. HMSO, London.

ROYAL COMMISSION ON ENVIRONMENTAL POLLUTION (1994). *Eighteenth report — transport and the environment*. HMSO, London.

SHERWOOD P.T. (1987). *Wastes for imported fill*. ICE Construction Works Guide. Thomas Telford, London.

SHERWOOD P.T. (1993). *Soil stabilization with cement and lime — a state of the art review*, HMSO, London.

SHERWOOD P.T. and ROE P.G. (1974). *The effect on the landscape of borrow-pits used in major roadworks*. Transport and Road Research Laboratory, Crowthorne, UK, Supplementary Report 122 UC.

SHERWOOD P.T. and RYLEY M.D. (1966). *The use of stabilized pulverized fuel ash in road construction*. Road Research Laboratory, Crowthorne, UK, Report LR 49.

SHERWOOD P.T. and RYLEY M.D. (1970). *The effect of sulphates in colliery shale on its use for roadmaking*. Road Research Laboratory, Crowthorne, UK, Report LR 324.

SHERWOOD P.T., TUBEY L.W. and ROE P.G. (1977). *The use of waste and low grade materials in road construction — 7. Miscellaneous wastes*. Transport and Road Research Laboratory, Crowthorne, UK, Report LR 819.

SWEERE G.T.H. (1989) (JONES R.H and DAWSON A.R. (eds)) Structural contribution of self-cementing granular bases to asphalt pavements. *Unbound aggregates in roads*. Butterworths, London. 343—353.

SWEERE G.T.H. (1991). Re-use of demolition waste in road construction. *Unbound Aggregates in Construction*, Nottingham.

TANFIELD D.A. (1978). The use of cement-stabilized colliery spoils in pavement construction. *International conference on the use of by-products and waste in civil engineering*, Paris.

THOMAS M.D.A., KETTLE R.J. and MORTON J.A. (1987). (RAINBOW A.K.M. (ed.)) Short-term durability of cement-stabilized minestone. *Reclamation, treatment and utilization of coal mining wastes*. Elsevier, Amsterdam, 533—544.

TUBEY L.W. (1978). *The use of waste and low-grade materials in road construction — 5. China clay sand*. Transport and Road Research Laboratory, Crowthorne, UK, Report LR 819.

TUBEY L.W. and WEBSTER D.C. *Effects of mica on the roadmaking properties of materials*. Transport and Road Research Laboratory, Crowthorne, UK, Supplementary Report 408.

VERHASSELT A. and CHOQUET F. (1989). (JONES R.H. and DAWSON A.R. (eds)) Steel slags as unbound aggregate in road construction: problems and recommendations. *Unbound aggregates in roads*. Butterworths, London, 204−209.

WATSON K.L. (1980). *Slate waste — engineering and environmental aspects*. Applied Science Publishers, London.

WEST G. and O'REILLY M.P. (1986). *An evaluation of unburnt colliery shale as fill for reinforced earth structures*. Transport and Road Research Laboratory, Crowthorne, UK, Research Report 97.

WHITBREAD M., MARSEY A. and TUNNELL C. (1991). *Occurrence and the utilisation of mineral and construction wastes*. Report for the Department of the Environment. HMSO, London.

WOOD C.E.J. (1988). A specification for lime-stabilization of subgrades. *Lime stabilization '88'*. BACMI, London, 2−8.